MW01479447

ENERGY CONSERVATION IN WATER AND WASTEWATER TREATMENT FACILITIES

Prepared by the **Energy Conservation in Water and Wastewater Treatment Facilities Task Force** of the **Water Environment Federation**

Ralph B. "Rusty" Schroedel, Jr., P.E., BCEE, *Chair*

Peter V. Cavagnaro, P.E., BCEE, *Vice-Chair*

Raul E. Aviles Jr., P.E., CPE, CEM, GBE
David M. Bagley, Ph.D., P.E.
Edward Baltutis
Glen R. Behrend, P.E.
Joseph Cantwell
Randall C. Chann, P.E.
S. Rao Chitikela, Ph.D., P.E., BCEE
John Christopher, P.E.
Stuart Kirkham Cole, Ph.D., P.E.
Peter R. Craan, P.E., CAP
Alex Ekster ,Ph.D., P.E.
Richard Finger
Eugenio Giraldo
Matthew J. Gray, P.E.
Mark R. Greene, Ph.D.
Tom Jenkins
Carl R Johnson, P.E., BCEE

Dimitri Katehis
Gregory Lampman
Lee A. Lundberg, P.E.
Venkatram Mahendraker, Ph.D., P. Eng.
James J. Marx, P.E.
Henryk Melcer
Indra N. Mitra, Ph.D., P.E.
Kathleen O'Connor, P.E.
Sudhanva Paranjape, P.E.
Vikram M. Pattarkine, Ph.D.
Marie-Laure Pellegrin, Ph.D.
Beth Petrillo
Mark Revilla, P.E.
David R. Rubin
Michael A. Sevener, P.E., BCEE
Brian D. Stitt
Don Voigt, P.E.
Thomas M. Walski, Ph.D., P.E., DAWRE
Roger C. Ward, P.E., BCEE
James E. Welp
Jianpeng Zhou, Ph.D., P.E., BCEE

Under the Direction of the **Municipal Design Subcommittee** of the **Technical Practice Committee**

2009

Water Environment Federation
601 Wythe Street
Alexandria, VA 22314–1994 USA
http://www.wef.org

ENERGY CONSERVATION IN WATER AND WASTEWATER TREATMENT FACILITIES

WEF Manual of Practice No. 32

Prepared by the Energy Conservation in Water and Wastewater Treatment Facilities Task Force of the Water Environment Federation

WEF Press

Water Environment Federation   Alexandria, Virginia

New York   Chicago   San Francisco   Lisbon   London   Madrid
Mexico City   Milan   New Delhi   San Juan   Seoul
Singapore   Sydney   Toronto

The **McGraw·Hill** Companies

Cataloging-in-Publication Data is on file with the Library of Congress.

McGraw-Hill books are available at special quantity discounts to use as premiums and sales pro-motions, or for use in corporate training programs. To contact a representative please e-mail us at bulksales@mcgraw-hill.com.

Energy Conservation in Water and Wastewater Treatment Facilities, MOP 32

Copyright © 2010 by the Water Environment Federation. All rights reserved. Except as permitted under the United States Copyright Act of 1976, no part of this publication may be reproduced or distributed in any form or by any means, or stored in a data base or retrieval system, without the prior written permission of WEF. Permission to copy must be obtained from WEF.

1 2 3 4 5 6 7 8 9 0 FGR/FGR 0 1 5 4 3 2 1 0 9

ISBN 978-0-07-166794-4
MHID 0-07-166794-6

The sponsoring editor for this book was Larry S. Hager and the production supervisor was Pamela A. Pelton. It was set in Palatino by Lone Wolf Enterprises, Ltd. The art director for the cover was Jeff Weeks.

Water Environment Research, WEF, and WEFTEC are registered trademarks of the Water Environment Federation.

This book is printed on acid-free paper.

IMPORTANT NOTICE

The material presented in this publication has been prepared in accordance with generally recog-nized engineering principles and practices and is for general information only. This information should not be used without first securing competent advice with respect to its suitability for any general or specific application.

The contents of this publication are not intended to be a standard of the Water Environment Feder-ation (WEF) and are not intended for use as a reference in purchase specifications, contracts, regu-lations, statutes, or any other legal document.

No reference made in this publication to any specific method, product, process, or service consti-tutes or implies an endorsement, recommendation, or warranty thereof by WEF.

WEF makes no representation or warranty of any kind, whether expressed or implied, concerning the accuracy, product, or process discussed in this publication and assumes no liability.

Anyone using this information assumes all liability arising from such use, including but not lim-ited to infringement of any patent or patents.

About WEF

Formed in 1928, the Water Environment Federation (WEF) is a not-for-profit technical and educational organization with 35,000 individual members and 75 affiliated Member Associations representing water quality professionals around the world. WEF and its Member Associations proudly work to achieve our mission of preserving and enhancing the global water environment.

For information on membership, publications, and conferences, contact

Water Environment Federation
601 Wythe Street
Alexandria, VA 22314-1994 USA
(703) 684-2400
http://www.wef.org

Manuals of Practice of the Water Environment Federation

The WEF Technical Practice Committee (formerly the Committee on Sewage and Industrial Wastes Practice of the Federation of Sewage and Industrial Wastes Associations) was created by the Federation Board of Control on October 11, 1941. The primary function of the Committee is to originate and produce, through appropriate subcommittees, special publications dealing with technical aspects of the broad interests of the Federation. These publications are intended to provide background information through a review of technical practices and detailed procedures that research and experience have shown to be functional and practical.

Water Environment Federation Technical Practice
Committee Control Group

R. Fernandez, *Chair*
J. A. Brown, *Vice-Chair*
B. G. Jones, *Past Chair*

A. Babatola
S. Innerebner
L. W. Casson
K. Conway
V. D'Amato
A. Ekster
R. C. Johnson
S. Moisio
T. Page-Bottorff
S. J. Passaro
R. C. Porter
E. P. Rothstein
A. T. Sandy
A. Tyagi
A. K. Umble
T. O. Williams

Contents

Chapter 3   Electric Motors and Transformers 37

Preface

Energy utilization and conservation are receiving increasing attention in the design and operation of wastewater and water facilities. The purpose of this manual is to help the manager, operator, and designer of municipal facilities identify and evaluate the processes and equipment which use the most energy, and then provide guidance and direction on the steps to be taken to reduce the energy consumption. Approaches to organization, measurement, and management of an energy savings program are also discussed.

This Manual of Practice was produced under the direction of Ralph B. Schroedel, Jr., P.E., BCEE, *Chair*, and Peter V. Cavagnaro, P.E., BCEE, *Vice-Chair*.

The principal authors of this Manual of Practice are as follows:

Chapter 1	Peter V. Cavagnaro, P.E., BCEE
Chapter 2	Roger C. Ward, P.E., BCEE
Chapter 3	Edward Baltutis
Chapter 4	Thomas M. Walski, Ph.D., P.E., DAWRE
Chapter 5	Peter R. Craan, P.E., CAP
Chapter 6	Sudhanva Paranjape
Chapter 7	Brian D. Stitt
Chapter 8	Vikram M. Pattarkine, Ph.D.
Chapter 9	Tom Jenkins
Chapter 10	Eugenio Giraldo
Chapter 11	Joseph Cantwell
Appendix A	Peter V. Cavagnaro, P.E., BCEE
Appendix B	Peter V. Cavagnaro, P.E., BCEE
Appendix C	Peter V. Cavagnaro, P.E., BCEE
Appendix D	Peter V. Cavagnaro, P.E., BCEE

Contributing authors and the chapters to which they contributed are David Fenster (2), John Christopher (7), Dimitri Katehis (7), Venkatram Mahendraker, Ph.D., P.Eng. (8), and David R. Rubin (8).

In addition to the WEF Task Force and Technical Practice Committee Control Group members, reviewers include Patrick L. Daigle, Tim Dobyns, Christopher Godlove, and Amy Santos.

Authors' and reviewers' efforts were supported by the following organizations:

AECOM, Lake Forest, California and Sheboygan, Wisconsin
AquaTEC Inc., Houston, Texas
Bentley Systems, Inc., Watertown, Connecticut
BioChem Technology, Inc., King of Prussia, Pennsylvania
Black & Veatch, Cincinnati, Ohio
Brown and Caldwell, Seattle, Washington
Carollo Engineers, Winter Park, Florida
CDM, Chicago, Illinois and Phoenix, Arizona
CH2M HILL, Chantilly, Virginia, and New York, New York
Clinton Foundation, Washington, DC
County Sanitation Districts of Los Angeles County, Whittier, California
Environmental Dynamics Inc., Columbia, Missouri
Ekster and Associates, Fremont, California
Energenecs Applications Engineering Group, Cedarburg, Wisconsin
ERM, Exton, Pennsylvania
ESCOR, Milwaukee, Wisconsin
Focus on Energy, Madison, Wisconsin
Georgia EPD, Atlanta, Georgia
GE Water & Process Technologies, Oakville, Ontario, Canada
Hazen and Sawyer, New York, New York
HDR, Tampa, Florida
HNTB Corporation, Indianapolis, Indiana
Johnson Controls, Inc., Milwaukee, Wisconsin
KCI Technologies, Sparks, Maryland
Malcolm Pirnie, Inc., White Plains, New York
McKim & Creed, Virginia Beach, Virginia
Metcalf & Eddy, Sunrise, Florida

New York State Energy Research and Development Authority (NYSERDA), Albany, New York

O'Brien & Gere Engineers, Syracuse, New York

OTT North America, Suwanee, Georgia

Red Oak Consulting, Fair Lawn, New Jersey

R. W. Beck, Inc., San Diego, California

Science Applications International Corporation, San Diego, California

Southern Illinois University Edwardsville, Edwardsville, Illinois

Turblex, Inc., Springfield, Missouri

University of Wyoming, Laramie, Wyoming

Veolia Water North America, Moon Township, Pennsylvania

List of Figures

Figure   **Page**

List of Tables

ENERGY CONSERVATION IN WATER AND WASTEWATER TREATMENT FACILITIES

Chapter 1

Energy Efficiency

1.0 INTRODUCTION

This manual addresses energy and how it is used in the operation of municipal water and wastewater facilities. Information provided for pumping at water and wastewater treatment plants (WWTPs) will also apply to raw water pump stations, low- and high-service water pumps, water booster pumps, and wastewater lift stations found outside the treatment plant fence line.

Water and wastewater facilities are typically among the community's largest energy consumers (California Energy Commission, 1990), accounting for 30 to 60% of

1

a municipal government's energy usage (U.S. EPA, 2008) and 3 to 4% of the nation's total energy usage.

Energy consumption is a significant operating cost at a treatment plant, consuming 15 to 30% of the operation and maintenance (O&M) budgets at a large WWTP and 30 to 40% at a small WWTP. Energy costs of operating these facilities continue to rise because of the trend in the cost of fuels, inflation, and increasing wastewater discharge requirements that result in the application of energy-intensive treatment processes.

Advancing technology offers new opportunities for improving energy efficiency and the application of new energy conservation measures (ECMs) at a treatment facility. *Energy conservation measures* are defined as a physical improvement, plant operation, or equipment maintenance practice that results in a reduction in utility or operating cost. Energy conservation measures are identified throughout this manual, as information about specific items of equipment and/or processes is discussed.

The manual includes basic concepts of energy and describes the technical basis for the energy required by unit operations and unit processes commonly found in water and wastewater facilities. A substantial amount of energy is also used in building- and facility-related equipment such as lighting and heating and ventilation. These subjects are already covered in references on energy conservation, some of which are listed in the section titled "Suggested Readings" at the end of this chapter.

The purpose of the manual is to serve as a primer for helpful suggestions for controlling energy and energy costs at a water or wastewater facility. It is intended for managers, plant engineers, and senior operators to provide a basis for greater understanding of the concepts of energy and efficient energy use to better manage consumption and demand. It will also be useful to the design engineer in selecting energy-efficient equipment and processes.

Each of the chapters provides a discussion of basic principles and concepts of energy requirements and potential sources of inefficiency. Background information will aid in the selection of equipment during design of new facilities or the selection of replacement equipment for an existing process.

Chapter 1—"Energy Efficiency," provides an overview of the manual and includes basic technical concepts of energy describing the forms and sources of energy. This chapter also includes a discussion on the relationship between energy and climate change.

Chapter 2—"Utility Billing Procedures and Incentives," describes how utility companies charge for their services and presents details on the various components of a utility bill. This chapter focuses on the overall costs of energy, and utility rate structures are explained in understandable terms.

Chapter 3—"Electric Motors and Transformers," describes the types of electric motors, the importance of matching motors to the connected load, and use of premium efficiency motors.

Chapter 4—"Pumps," presents information on the principles associated with the capacity of pumps, determining the amount of energy used by pumps, and factors affecting power consumption.

Chapter 5—"Variable Controls," describes the types of adjustable controls, adjustable frequency drives and factors considered in their application, and factors to consider in the control of pumps and blowers.

Chapter 6—"Energy Utilization in Water Treatment Processes," discusses the electricity-consuming equipment and systems found in water treatment plants such as motors, pumps, membranes, and UV disinfection systems.

Chapter 7—"Energy Utilization in Wastewater Treatment Processes," discusses the electricity-consuming equipment and systems found in WWTPs: motors, pumps, and aeration systems.

Chapter 8—"Wastewater Aeration Systems," describes how oxygen requirements are determined, types of aeration equipment available, factors affecting design and oxygen transfer efficiency, operational considerations, and opportunities for energy conservation.

Chapter 9—"Blowers," explains the importance of blowers to the total power consumed at a WWTP, considerations for blower applications, factors in the operation of positive-displacement blowers, operating principles of centrifugal blowers, and retrofit opportunities.

Chapter 10—"Wastewater Solids Processes," is devoted entirely to solids-handling processes and discussions on energy recovery from anaerobic digestion systems and controlling energy losses from incinerators and sludge dryers.

Chapter 11—"Energy Management," presents ways to control electrical demand such as how costs can be reduced through a better understanding of energy consumption, utility rate structures, and effective energy management. Ways to use alternate energy sources for greater overall economy are addressed. Examples are shown to demonstrate how to compare differences in initial capital cost with operating cost changes because of differences in equipment efficiency.

2.0 ROLE OF MANAGEMENT

The overarching responsibility of a water facility is to protect public health and, for a wastewater treatment facility, to meet discharge limits. However, water and wastewater management responsibilities include establishing long-term objectives; developing supporting plans to accomplish the objectives; providing work direction, staff selection, and staff development; and ensuring overall facility performance.

Management sets goals for energy management and the associated programs should be defined, monitored, executed, and ongoing. Committed management provides leadership, encouragement, and resources to conserve energy.

2.1 Monitoring

The plant cannot manage what is not measured and, as such, supervisory control and data acquisition (SCADA) systems should monitor and report kilowatts and joules (kilowatt-hours) as well as key indicators (joules per liter [J/L] for pumps and joules per cubic meter [J/m^3] for blowers) to provide meaningful measures of production compared to energy demand and consumption. The energy rate schedule(s) should be reviewed to determine if there are time-of-use charges and how "demand" charges are established when programming the output reports of the SCADA system. Alternatively, power monitoring equipment can be installed on key pieces of equipment so that staff has the information needed to assess the impact of operational decisions on utility bills.

2.2 Resources

A number of on-site resources are commonly available at a water or wastewater facility. These include engineering drawings that are a primary source of information regarding the locations, elevations, and details of installation, which will be useful in locating larger motors; assessing the total dynamic head on a pump; and providing information related to the aeration blowers. One-line diagrams provide insight to the location and size of motors. The hydraulic profile provides useful insight to energy requirements for pumping. Construction shop drawings provide accurate nameplate data for installed motors, pumps, blowers, and other equipment. Additionally, the O&M manual contains basic design data and often provides information for the design intent and control logic.

Operating data are used to establish baseline flows and loads to the plant. Daily, monthly, and annual average values for biochemical oxygen demand and ammonia nitrogen provide the information needed to assess aeration requirements. The method for using this data to compute aeration requirements is presented in Chapter 8 ("Aeration Systems").

Operation and maintenance manuals serve as information resources and should be kept up-to-date to document energy conservation features designed or added to a WWTP. For example, if adjustable-frequency drives (AFDs) have been installed on a pump as an ECM, the manual should provide information that the operator or supervisor needs on the particulars of how to monitor the system's energy use and how to knowledgeably operate AFDs to minimize energy consumption. Data on pump system efficiency at various operating conditions should be presented on curves and charts in the manuals that directly relate to information on field gauges, monitors, and computers. The manual should explain the design and provide basic design criteria. Operation and maintenance manuals should also be updated to include any operating improvements developed and implemented by staff or consultants.

Energy bills and other records of energy consumption for electricity, fuel, gas, and chemicals should be accessible to plant management for review. Staff should be made aware of factors affecting energy bills, the level of consumption, and costs. Lead operators should be aware of where the electric meter and submeters are as well as how to read the meter. The manufacturer's Web site commonly contains technical data sheets for the meter.

The U.S. Environmental Protection Agency (U.S. EPA) has published information on energy conservation in WWTPs (see the "Suggested Readings" section at the end of this chapter). Their Sustainable Infrastructure for Water & Wastewater Web site (http://www.epa.gov/waterinfrastructure/bettermanagement_energy.html) also provides useful information on energy conservation. State agencies such as the New York State Energy Research and Development Agency (NYSERDA), the California Energy Commission (CEC), and Wisconsin's Focus on Energy also provide unique and valuable information based on years of practical experience. In addition, the local utility's Web site should be visited for insights on resources and the availability of funding.

Professional societies such as the Water Environment Federation (WEF), Water Environment Research Foundation, American Water Works Association, Water Research Foundation, and the Association of Energy Engineers all have resources related to energy conservation in water and wastewater facilities. Monthly newsletters, magazines, journals, and conference proceedings are all rich sources of information.

A list of energy-related agencies and organizations is presented in Appendix A.

3.0 ENERGY AND POWER

The term, *energy*, has different interpretations. In this manual, *energy* is defined by the science of physics and is the ability or capacity to do work.

There are many different types of energy. These different types can be divided into two broad classifications: potential energy and kinetic energy. Kinetic energy is possessed by a moving object. Potential energy is stored and contained within fuel oil or in an elevated water tank or reservoir. Potential energy is converted to kinetic energy when released. Energy can be changed from one form to another. Common forms of energy are chemical, electrical, mechanical, thermal, radiant, and nuclear. Most people are familiar with the conversion of fossil fuel through burning (chemical energy) to form steam (thermal energy).

The most common form of energy used in water and wastewater facilities is electric energy. Natural gas is also commonly used as fuel in boilers to provide comfort and process heat. Other forms of energy that may be used at a treatment plant site are fuel oil, propane, and steam.

Power is the rate at which work is performed. It is the speed at which the work is done or the rate at which energy is expended. The basic unit of power in the metric system is the watt, which gives rise to the production of energy at the rate of one joule per second (J/s). In the English system, the common units of measure for power are horsepower (1 hp = 550 ft-lb/sec) and British thermal units per hour (Btu/hr). Note that watt and horsepower do not appear to be rate units, but they are. All power units are measured as work per unit time.

The size of a motor or engine is defined not by the total amount of work to be done, but by the rate at which it can do work. Electric motor power is commonly measured in kilowatts (horsepower), boiler output is measured in watts (Btu/hr or boiler horsepower = 33 520 Btu/hr), and air-conditioner power is measured in watts (Btu/hr or tons of refrigeration = 12 000 Btu/hr).

The power applied multiplied by the time it is applied represents total energy consumption, which is expressed as joules (J) or kilojoules (kJ). Energy consumption is typically measured in joules (kilowatt-hours [kWh]) if consumed as electricity and in kilojoules (British thermal units [Btu]) if consumed as a fuel.

Energy (joule) = power (kW) × time (hr)
Energy (hp-hr) = power (hp) × time (hr)
Energy (Btu) = power (Btu/hr) × time (hr)
Energy (ft-lb) = power (ft-lb/sec) × time (sec)

TABLE 1.1   Conversion chart for various forms and units of energy

Unit	Work		Heat		Electric	
	ft-lb	kg-m	Btu	kcal	hp-hr	kWh
ft-lb	1	0.138 3	1.286×10^{-3}	3.241×10^{-4}	5.050×10^{-7}	3.766×10^{-7}
kg-m	7.231	1	9.302×10^{-3}	2.343×10^{-3}	3.654×10^{-6}	2.724×10^{-6}
Btu	777.9	107.5	1	0.252 0	3.927×10^{-4}	2.928×10^{-4}
kcal	3 086	426.8	3.968	1	1.558×10^{-3}	1.162×10^{-3}
hp-hr	1.980×10^{6}	2.737×10^{5}	2 547	641.7	1	0.745 7
kWh	2.655×10^{6}	3.671×10^{5}	3 415	860.5	1.341	1

Note: kWh $\times 2.78 \times 10^{-7}$ = joules

Measured at standard temperature (15.56 °C [60 °F]) and pressure (101.56 kPa [14.73 psia]), natural gas contains approximately 37 000 kJ/m³ (1000 Btu/cu ft), and billing units are typically 2.8 cubic meters (100 cu ft) of gas, 28 m³ (1000 cu ft or dekatherm), 105 000 kJ (100 000 Btu, otherwise known as *therms*), or million Btu (MMBTU).

Fuel oil characteristically contains 3.9×10^{4} kJ/L (140 000 Btu/gal), and consumption is measured in liters (gallons). Table 1.1 provides conversion factors so that energy consumption measured by one method may be changed to different units. In addition, Appendix B provides metric conversions for English units used in this manual.

4.0   CLIMATE CHANGE

4.1   Water Environment Federation Resolution on Climate Change

The Board of Trustees of WEF (2006) charged WEF Committees to identify and disseminate information that contributes to the reduction of greenhouse gas (GHG) emissions from wastewater facilities. In the same resolution, WEF urged municipal wastewater treatment agencies to become leaders in their communities for reducing and mitigating climate change.

In the course of treating wastewater and handling and disposing of wastewater solids, municipal wastewater facilities contribute to the emission of GHGs. This section briefly describes what the gases are, provides examples of emission sources, and discusses the relationship of GHGs to the carbon footprint. The reader is pointed to sources of information for computing GHG emissions.

4.2   Greenhouse Gas Overview

Greenhouse gases trap heat in the atmosphere. Some GHG emissions occur naturally, whereas others are the result of human activity. The principal gases entering the atmosphere as a result of human activities are carbon dioxide (CO_2), methane (CH_4), nitrous oxide (N_2O), and fluorinated gases (visit U.S. EPA's Climate Change-Greenhouse Gas Emissions Web site at http://www.epa.gov/climatechange/emissions/index.html#proj).

Carbon dioxide enters the atmosphere when fossil fuels are burned. It is removed (or sequestered) when it is absorbed by plants. Carbon dioxide represents about 77% of global GHG emissions (World Resources Institute, 2006). Methane results from the decay of organic wastes, such as that that occurs in a municipal solid waste landfill, originates from certain agricultural and livestock operations, and also results from the mining and handling of carbon-based fuels. Methane represents 14% of global GHG emissions. Nitrous oxide is emitted during combustion of fossil fuels and from certain agricultural and industrial activities, and represents 8% of global GHG emissions. Fluorinated gases are synthetic gases emitted from industrial processes, and represent 1% of the GHG emissions.

Greenhouse gases vary in their ability to trap heat in the atmosphere (World Resources Institute, 2006). Each gas has a *global warming potential,* which refers to the potential to trap heat relative to that of carbon dioxide. Therefore, GHGs are commonly reported in terms of the equivalent amount of carbon dioxide or carbon dioxide equivalents (CO_2 e). The sum of GHGs produced, measured as CO_2 e within a specified boundary and for a specified period of time, is the carbon footprint.

4.3   Sources of Greenhouse Gas Emissions

This section highlights sources of GHG emissions, although it does not represent a comprehensive list. Water and wastewater facilities are significant consumers of electric power, much of which is generated by burning carbon-based fuels. Emissions of carbon dioxide and nitrous oxides at power generation plants contribute to the carbon footprint at water or wastewater facilities.

The oxidation of organics to carbon dioxide and water in wastewater biological treatment will result in emissions of CO_2. Other sources of GHG emissions are from fuel oil used to operate standby generators and fuel used to power trucks that are used to transport biosolids for disposal. Disposal of solid waste will also have GHG emissions associated with it.

Wastewater facilities that have lagoons or sludge storage ponds will result in the release of carbon dioxide and may release methane from the decomposition of the organic wastes. These emissions will contribute to the facilities' carbon footprint. Likewise, byproducts produced in an anaerobic sludge digester create GHG emissions, directly as methane and carbon dioxide, and as carbon dioxide and nitrous oxide if burned in a flare or process boiler, or as carbon dioxide and nitrous oxide if burned in a turbine or internal combustion engine to generate electricity or to power a pump or blower.

Advanced treatment processes require relatively large amounts of energy, much of which is embodied in chemicals added for purposes such as phosphorus precipitation, refractory organic removal, and desalinization. The nitrous oxide formed by and emitted from biological nutrient removal processes is an area of concern because of the CO_2 of this gas. Membrane processes being incorporated into water and WWTPs will most likely increase the energy intensity of the plant. For the purpose of computing carbon footprint, consideration may need to be given to the amount of energy embodied in producing and transporting the chemicals.

The operation of pumps and blowers consumes a significant amount of electricity at WWTPs and most of the energy at water treatment plants. Conserving this electricity will reduce the amount of power needing to be generated by the electric power utility, thereby reducing emissions of GHGs. Water and WWTPs have facilities that must be lighted, and using up-to-date lighting can reduce consumption. The same principle applies to heating, which uses natural gas, electricity, and/or propane.

4.4 Computing Greenhouse Gas Emissions

Computation of GHG emissions is beyond the scope of this Manual of Practice. However, the following information is provided to offer a source of resources that the reader may pursue to compute emissions.

The Intergovernmental Panel on Climate Change (IPCC) (2006) is commonly referred to as a source of national guidelines for computing GHG emissions. The IPCC guidelines account for only methane and nitrous oxide, and do not count carbon dioxide emissions.

A procedure specific to the wastewater industry described by Monteith et al. (2005) was evaluated using full-scale data from 16 WWTPs in Canada and was applied to plants in all Canadian provinces. For aerobic processes primarily used in North America, the authors determined that the principal GHG emitted from municipal WWTPs was carbon dioxide.

Additional insights may be obtained from the Greenhouse Gas Protocol Initiative (http://www.ghgprotocol.org/), which was jointly organized in 2008 by the World Business Council for Sustainable Development and the World Resources Institute. The Greenhouse Gas Protocol is a widely used international accounting tool for understanding, quantifying, and managing GHG emissions.

Another potential resource for accounting for GHG emissions on an enterprise-wide basis is the U.S. EPA Climate Leaders Web site (http://www.epa.gov/climate-leaders). This system, which is often used in the United States, assigns a different emissions factor for each of 26 U.S. regions; every postal zip code is then mapped to one of those regions.

Another source of information that offers a number of calculation tools for expressing reductions in terms of the number of cars, barrels of oil, trees, homes, and so on is the U.S. EPA Clean Energy Web site (http://www.epa.gov/cleanenergy/energy-resources/refs.html).

Insight on emissions avoided from peaking power plants that are typically dispatched during periods of highest demand and that which are typically the least efficient can be obtained from the Cleaner and Greener Calculators Web site at http://www.cleanerandgreener.org/resources/calculators.htm. (Cleaner and Greener is a program of the Leonardo Academy, Inc., Madison, Wisconsin.)

5.0   IDENTIFYING ENERGY CONSERVATION MEASURES

An ECM should either reduce the amount of energy consumed or reduce the price paid for energy while maintaining a reasonable amount of reliability to meet discharge limits. Note that it is possible to reduce the amount of energy consumed without a proportional reduction in the billed amount. Demand charges reflect a significant portion of the electric bill and may not be affected by conservation. The operator should review the rate schedule as indicated on the bill or obtain a description of the rate schedule, which is commonly available on the Web site for the local utility. The reader should refer to Chapter 2 for further explanation.

Simple steps for identifying ECMs are as follows:

1. Understand and become familiar with the procedures used by the power company to bill for energy used.
2. Analyze utility bills and request time-of-day charts for kilowatts and kilowatt-hours used at the WWTP. Review the trend for demand charges and

the opportunity to reduce demand by changing the schedule for elective operations, such as exercising equipment with large motors during periods of off-peak demand charge. Identify charges related to reactive power and the power factor.

3. Review historical energy consumption (36 months of data is preferred, but an analysis can be performed with 24 months of data as a minimum) by the WWTP as a whole as well as any components that are individually metered.

4. Develop a list of significant energy-consuming equipment. Begin with units with the largest connected horsepower first, and then factor based on the amount of time that the equipment is operated. Performance tests can provide useful information for power draw and equipment efficiency. The U.S. Department of Energy's Industrial Technologies Program has a Pumping System Assessment Tool (http://www1.eere.energy.gov/industry/best-practices/software.html#psat) that can be used to evaluate raw and treated water pumps. Consideration should also be given to large rooftop-mounted heating, ventilation, and air conditioning (HVAC) equipment.

5. Define current process control procedures.

6. Analyze data collected to identify equipment presenting the greatest opportunity for improvement. Research the unit and identify potential ECMs.

It is useful to focus the initial search for ECMs in those areas where the greatest amounts of energy are consumed and, therefore, the greatest potential for savings exists. Most often, aeration is the single largest energy user (commonly around 50% of the total), followed by pumping systems. At some WWTPs, other systems can use a large percentage of the electricity or energy. Energy conservation measures such as use of premium efficiency motors are focused on energy conservation. The following are ECMs to consider:

- Connections and switches on all major WWTP power-driven equipment, buses, and transformers should be checked at least once per year so that simple corrective action can be taken before serious problems with equipment develop or significant power losses occur.

- Motors should be operated as close to nameplate voltage as practical because any deviation from the nameplate rating affects the motor's efficiency. In general, it is recommended that the motor line drop not exceed 5% of the line voltage.

Benchmarking is the comparison of a WWTP's overall energy usage to published typical uses of electricity, fuel, or chemicals at other WWTPs. A simple but effective gauge is to determine the unit energy consumption in joules per cubic meter (J/m³) or kilowatt-hours per million gallons (kWh/mil. gal) treated. Figure 1.1 summarizes typical energy consumption at a WWTP in the western United States. As indicated, lagoon, trickling filter, and rotating biological contactor WWTPs are the most efficient, with the activated sludge process and oxidation ditches being large energy consumers. Figure 1.1 assumes that the WWTPs have influent pumping, primary clarification, anaerobic sludge digestion (except oxidation ditches), no effluent pumping, and no cogeneration.

In October 2007, the U.S. EPA Energy Star Program initiated the first online benchmark for WWTPs (visit U.S. EPA's Energy Star Portfolio Manager Web site at http://www.energystar.gov/istar/pmpam/). In January 2008, a similar benchmark was initiated for water treatment plants. The benchmarks are based on statistical

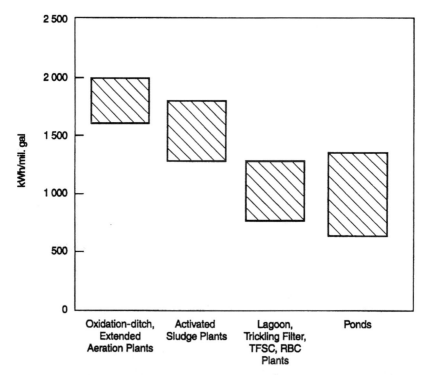

FIGURE 1.1 Typical unit energy consumption of various types of wastewater treatment plants (TFSC = trickling filter solids contact and RBC = rotating biological contractor) (kWh/mil. gal × 951.1 = J/m³).

analysis of the survey data and models established for estimating benchmark levels based on size, treatment process (if applicable), and location.

The 1997 edition of this manual of practice presented four tables that have served as a guide to computing energy consumption in WWTPs. Estimates of electricity used in each WWTP unit process was presented for four categories of plants: trickling filters, activated sludge, advanced WWTPs with nitrification, and advanced WWTPs without nitrification. These tables have been used by plants to establish baseline conditions and continued access to this data was determined to be beneficial. Therefore, these tables have been retained and appear in Appendix C of this manual. A line has been added to the top of each table for units of flow from the International System of Units. The actual energy used will vary at each WWTP, therefore, energy usages in these tables should be adjusted accordingly. Adjustments should be made for site-specific conditions and differences in treatment processes such as odor control, intermediate pumping, high-purity oxygen, biological nutrient removal, enhanced nutrient removal, membrane processes, ultraviolet disinfection, water reuse pump stations, gravity belt or drum thickening, and centrifuge dewatering. Information has been added at the bottom of each table to provide assistance with modifications for ultraviolet disinfection and gravity belt thickeners. The power consumed by the other processes is affected by a number of variables and information should be requested from the vendor and engineers to establish the proper level of energy intensity.

Measurement and verification (M&V) refers to accounting for the savings associated with each ECM. Measurement and verification requires establishment of an initial baseline by measurements or other means, and conducting follow-up measurement or other accounting. Measurement and verification strategies are described in the International Performance Measurement and Verification Protocol (Efficiency Valuation Organization, 2007). Although these strategies are straightforward, they require some explanation, which is provided in the manual. The U.S. Department of Energy has defined M&V requirements for federal energy projects (U.S. Department of Energy, 2000).

Baseline measurements refer to the analysis of existing energy bills and operating data to identify the current level of consumption, peak energy usage, and costs for an existing water/wastewater facility, process, or system. Baseline measurements are made before implementing any ECMs so that the positive effect of each ECM can be measured.

Plant-wide baseline measurements are made by reviewing historical energy bills, based on the billing meter(s) where power enters the plant. A plant that has cogeneration must remember to include the energy measurements for the cogeneration facility. If a generator is being used to meet peaking demands, then that energy must be metered as well. Many plants are implementing photovoltaic solar, which must be accounted for as well.

The baseline for individual pieces of equipment is established by performing data logging for a period of time to pick up hourly, daily, and weekly variations. Data logging should be performed by qualified technicians who are experienced in installing the sensors and analyzing the data. Appendix D provides insight on the relationship between voltage, amperage, and power.

The savings associated with some improvements may not justify on-site monitoring, and mathematical computations (models) may be used to determine pre-savings and post-savings based on spot measurements and rules of thumb used in the industry. *Post-measurements* are made in order to compare results of the improvements.

Field measurement is necessary when conducting energy studies, establishing energy baselines, or conducting post-measurements. The correct method for measuring energy inputs is affected by a number of variables that depend on the conditions at the site. Field measurement requires the proper equipment, knowledge, experience, and safety procedures according to the location, power supply, and type of equipment. For example, lighting and power outlets are typically supplied with 120-V power; motors that operate process equipment are commonly powered by 480-V, three-phase, alternating current power, although some plants are served by 208-V three-phase power or some equipment by a 4160-V power supply. No one should work with electrical equipment unless they are experienced and trained to do so and possess the proper personal protective equipment.

6.0 RANKING AND IMPLEMENTATION OF ENERGY CONSERVATION MEASURES

Proper ranking of ECMs is important to the success of an energy conservation program. The goal of the program is to reduce consumption and costs while maintaining a reasonable degree of reliability of plant performance. Other factors to consider include management time to implement, decreases and increases in O&M costs, the time required to recover this cost, and the probability that a particular ECM will achieve its projected savings within the desired time frame.

Energy conservation measures may be divided into three categories based on cost and the degree of difficulty in terms of implementation: low cost and relatively easy to implement; moderate cost, requiring some technical analysis; and high cost, requiring field measurements and engineering analysis to implement. The actual dollar amounts assigned to these categories will vary with WWTP size and financial resources.

The first group includes actions requiring change in operational routine and offers a high degree of value. Typically, the funds needed for implementation will come from an existing WWTP expense budget. As a rule, these ECMs should be implemented at the earliest possible date.

The moderate-cost group may require staff evaluations, minor drawings or specifications, outside contractor assistance, and budget item procurement of labor and materials. Energy conservation measures in this group can be readily compared and prioritized on the basis of simple payback periods. The simple payback period is calculated by dividing the cost of implementing the ECM by the net annual savings (reduced O&M costs from existing system minus new O&M costs for the proposed ECM) it will provide. Energy conservation measures with the shortest payback periods are usually given the highest priority.

The last group of ECMs requires evaluation by the engineering staff and/or an outside engineering consultant. Evaluation and implementation would probably be separate budget line items. Major structural rehabilitation, automation systems, major equipment change-outs, and changes related to the treatment process are examples of actions in the group. Economic analyses that take into account the time value of money are required to compare alternative high-cost ECMs and justify their implementation.

7.0 CASE STUDIES

7.1 California Energy Commission

The California Energy Commission's (CEC's) energy efficiency programs provide technical assistance and low interest loans to help make water and wastewater treatment facilities more energy-efficient. The CEC's Energy Partnership Program also provides assistance with energy efficient design of new facilities. Recent projects include installation of premium efficiency motors, variable frequency drives, premium efficiency pumps, aeration diffuser upgrade, cogeneration, replacement of variable frequency drives, dissolved oxygen process control, blower replacement, and anoxic/anaerobic selectors.

Under a peak load reduction program (California Energy Commission, 2004), water and wastewater facilities reduced more than 52 MW of demand at a cost of less than $5 million to the Energy Commission in the form of grants ($250 to $300 per peak kilowatt reduced) in the following four categories: distributed generation, energy efficiency, load shedding, and curtailment. Examples of the distributed generation projects included refurbishment of an internal combustion cogeneration system using digester gas, installation of micro turbines to run on digester gas, and conversion of equipment to run on a blend of fuels including digester gas. Examples of the energy efficiency projects included codigestion of biosolids and food wastes, new aeration blowers, dissolved oxygen control, SCADA control, heat recovery, new motors, new pumps, storage and pump stations for off-peak pumping, and application of solar aerators.

7.2   Gloversville-Johnstown Joint Wastewater Treatment Facility

The communities of Gloversville, New York, and Johnstown, New York, operate a joint wastewater treatment facility (Ostapczuk, 2007) in Johnstown, New York. The plant was commissioned in 1972 and upgraded in 1992, has a permitted flow of 52.2 ML/d (13.8 mgd), and an average daily flow of 22.7 ML/d (6 mgd). There are 26 industrial users. In response to declining revenues from industrial customers and to avoid rate increases, the plant initiated a program to reduce energy costs. With assistance from NYSERDA, improvements were made to the aeration system in 2000 and to the anaerobic digester in 2002. An energy evaluation in 2005 resulted in installation of a dual membrane gas holder, improvement to the digester mixing and gas handling equipment, and provision for codigestion of biosolids with acid whey. The 2000 improvements to the aeration system included upgrade of aeration diffusers, a new aeration blower, and automated dissolved oxygen control.

8.0   REFERENCES

California Energy Commission (1990) The Second Report to the Legislature on Programs Funded Through Senate Bill 880; PL 400-89-006; Sacramento, California.

California Energy Commission (2004) SB 5X Water Agency Generation Retrofit Program - Final Report; Prepared by HDR Engineering, Folsom, CA:.

U.S. Department of Energy (2000) M&V Guidelines—Measurement and Verification for Federal Energy Projects; Version 2.2, DOE/GO-102000-0960; U.S. Department of Energy, Office of Energy Efficiency and Renewable Energy, Federal Energy Management Program. Efficiency Valuation Organization (2007) International Performance Measurement and Verification Protocol,

Concepts and Options for Determining Energy and Water Savings, Vol. 1, April; EVO 10000-1.2007. http://www.evo-world.org (accessed Feb 2008).

Intergovernmental Panel on Climate Change (2006) 2006 IPCC Guidelines for National Greenhouse Gas Inventories. http://www.ipcc-nggip.iges.or.jp/public/2006gl/index.htm (accessed Feb 2009).

Monteith, H. D.; Sahely, H. R.; MacLean, H. L.; Bagley, D. M. (2005) A Rational Procedure for Estimation of Greenhouse-Gas Emissions from Municipal Wastewater Treatment Plants. *Water Environ. Res.,* **77** (4).

Ostapczuk, R. (2007) Gloversville-Johnstown Joint Wastewater Treatment Facility Energy Conservation Program Case Study. *Proceedings of the 80th Annual Water Environment Federation Technical Exhibition and Conference* [CD-ROM]; San Diego, California, Oct 13–17; Water Environment Federation: Alexandria, Virginia.

U.S. Environmental Protection Agency (2008) *Ensuring a Sustainable Future: An Energy Management Guidebook for Wastewater and Water Utilities;* GS-10F-0337M; U.S. Environmental Protection Agency, Office of Wastewater Management. http://www.epa.gov/waterinfrastructure/pdfs/guidebook_si_energymanagement.pdf (accessed Feb 2009).

Water Environment Federation (2006) Resolution on Climate Change, Adopted in Dallas, Texas, on October 20, 2006. http://www.wef.org/GovernmentAffairs/Policy-PositionStatements/ClimateChange.htm (accessed Feb 2009).

World Resources Institute (2006) Hot Climate, Cool Commerce: A Service Sector Guide to Greenhouse Gas Management; World Resources Institute: Washington, D.C.

9.0 SUGGESTED READINGS

Dranetz BMI (1991) Dranetz Field Handbook for Electrical Energy Management; Dranetz BMI: Edison, New Jersey.

Toliyat, H. A.; Kliman, G. B. (2004) Handbook of Electric Motors; CRC Press, Taylor & Francis Group: London, England.

U.S. Environmental Protection Agency (2006) Wastewater Management Fact Sheet – Energy Conservation; EPA 832-F-06-024; U.S. Environmental Protection Agency, Office of Water, http://www.epa.gov/owm/mtb/energycon_fasht_final.pdf (accessed Feb 2009).

U.S. Environmental Protection Agency (1978) *Energy Conservation in Municipal Wastewater Treatment;* EPA-430/94-77-011; Washington, D.C.

U.S. Environmental Protection Agency (2008) Energy Star—Buildings & Plants Web Site. http://www.energystar.gov/index.cfm?c=industry.bus_industry_ elevating (accessed Feb 2009).

U.S. Environmental Protection Agency (1973) *Electrical Power Consumption for Municipal Wastewater Treatment;* EPA-R2/73-281; Washington, D.C.

Chapter 2

Utility Billing Procedures and Incentives

(continued)

1.0 INTRODUCTION

Understanding how a utility bill is calculated and how to read a utility meter can lead to successful reductions in utility costs. Utility rate structures have no direct effect on energy efficiency or energy use, although high charges may provide incentives to look for ways to reduce a wastewater treatment plant's (WWTP's) utility costs and to implement energy conservation measures. There are many opportunities to save money by understanding how current rate structure is being applied to current billing and knowing alternate billing structures that may be available. In addition to the standard industrial tariff, a variety of electric rate schedules and special contracts are commonly available that are based on the amount of electricity being consumed, the size of the provider, and regulations of the state in which the WWTP is located. Such special rates and contracts are designed to attain the load distribution and other management objectives of the utility serving the area. Additionally, the utility may be under federal or state mandates to encourage users to practice energy conservation or load-shifting that may reduce rates or provide rebates.

Deregulation of utilities may offer additional opportunities for large consumers who can take advantage of less expensive nonlocal suppliers. Like the deregulation of the phone company, natural gas is already being provided by nonlocal suppliers through local supplier transmission mains. Similarly, the Energy Policy Act of 1992 has opened up channels for supply of electricity by nonlocal providers at the wholesale level. Implementation at the retail level, known as *retail wheeling,* is up to each state because regulations must be approved by state utility regulatory commissions. Retail wheeling of electricity is not widely available. In this chapter, a broad overview of the types of utility charges is presented along with cost-saving opportunities offered by local utilities and key factors that can influence selection of an appropriate rate or contract option.

2.0 COST-SAVING OPPORTUNITIES

2.1 Obtain the Most Favorable Tariff

Electric utilities may have many different tariffs (rate schedules) that apply to a facility. For tariffs involving peak demand charges and differing time-of-day or seasonal consumption rates, power factor penalties/credits may or may not be favorable compared to flat rates based on consumption only. Additionally, the billing for an unused standby service may or may not be favorable compared to the life cycle cost of constructing, operating, and maintaining an on-site standby power-generating facility. A flat consumption rate tariff might be more appropriate for collection system lift stations because the electric loads are typically not controllable. Working with the electric utility to determine the most favorable tariff may result in significant cost savings.

2.2 Install High-Efficiency Transformers

The electric utility bill may incorporate the cost of the utility providing on-site medium voltage transformers (which, for example, reduce the utility's transmission voltage of 4160 V and higher to 480 V). Life cycle cost savings may be realized by purchasing and maintaining these transformers, particularly if they are a high-efficiency type.

2.3 Shave Peak Demand and Transfer Loads

If the tariff includes higher time-of-day demand charges and/or consumption rates, the cost of constructing, operating, and maintaining on-site power generation facilities to shave peak demand or reduce consumption could offset higher demand charges and consumption rates. Additionally, operating facilities such as sludge handling systems during off-peak periods (i.e., load transfer) could result in immediate savings.

2.4 Improve the Power Factor

If the tariff includes a penalty for a low power factor (e.g., less than 85%) or a credit for a near unity (greater than 98%) power factor, the cost of installing a capacitor bank for the entire plant or individual capacitors for the higher load motors may result in life cycle cost savings.

2.5 Install High-Efficiency Motors

The cost of installing high-efficiency motors on pumps and blowers that operate nearly continuously may be more than offset by the life cycle savings in the energy consumption and demand charges associated with those loads. Caution should be exercised when using high-efficiency motors because the higher starting current associated with high-efficiency motors may exceed the capacity of existing circuit protective devices (i.e., breakers and fuses).

3.0   ELECTRICITY

3.1   Billing Charges

Billing for electric service can take many forms depending on the economic objectives and administrative concerns being addressed by the utility. Unlike an electric invoice for home use, which may only contain a customer charge, energy charge, and taxes, WWTPs typically are subject to complex rate structures classified under the heading "industrial use." Although municipalities will often have rate classifications separate from industry, they are nonetheless similar to industrial rate classifications. Typically, the monthly invoice for all industrial consumers will contain the following minimum charge categories: customer, energy (or consumption), demand, power factor penalty or credit, fuel-cost adjustment, and taxes (note that most municipal WWTPs are exempt from state and local taxes). Additionally, there are many other charges and surcharges that may be added to the bill or credits given that may be deducted. Rates for energy and demand may be specified in use brackets.

3.1.1   Customer

The customer charge is a fee structured to compensate the utility for administrative costs incurred in servicing the customer (e.g., for reading meters, preparing and mailing bills, and collections). Customer charges for WWTPs are lower than energy and demand charges. Specifically, they are often less than $100 per month, but can be higher depending on the sophistication of the meter used.

3.1.2   Energy

The charge for energy (consumption) is based on the actual energy used, measured in kilowatt-hours (kWh) or megajoules (MJ), during the billing period. It represents the utility company's operating cost for generating and supplying the electricity, including its profits.

Utilities may assess premium charges for energy during peak usage periods to discourage energy use during these peak periods. Alternatively, lower costs for non-peak periods are also offered to encourage users to shift the time of their usage. This is a utility's effort to balance its production schedules and maximize use of existing facilities. Peak periods of usage in many parts of the United States occur during the hottest days of the summer, when air conditioning demand is greatest. Utilities must strain their resources to deliver the energy demanded by their customers during these brief peak demand periods, and may require the production of power by smaller natural gas-fueled peaking generators and the purchase of power on the open market. This condition may also be reflected in the demand charge discussed in the following section.

Energy use is measured using a wattmeter, which is an instrument for measuring the electric power (or the supply of electrical energy) in watts by concurrently measuring voltage and current. Older units are electromechanical, with the familiar hard-to-read clock-type dials, whereas modern units are electronic and have digital readouts.

3.1.3 *Demand*

Demand is the maximum power drawn during the billing period (usually averaged over a contiguous 15-minute or longer period) and is the combined power of all motors and other electrical devices in use during the facility's period of greatest electrical requirement. It represents the average sum of the power drawn in kilowatts for each motor and each electrical device that is on or running during the peak 15-minute period.

Demand is measured by the wattmeter, which integrates or sums up energy usage over the demand interval. It operates like a continuous moving average, whereby it drops off the first minute of use while adding the last minute of use during the interval. Recall from Chapter 1 that power = energy/time, and that power is measured in kilowatts and energy in megajoules (kilowatt-hours). The demand meter actually counts kilowatt-hours (megajoules) used during the interval. For example, if the wattmeter counted 2000 MJ (550 kWh) of energy use during a 15-minute demand interval, it would report 2200 kW of demand (550 kWh/0.25 h). Electromechanical units typically have a ratchet mechanism whereby each new peak demand value notches the pointer up to the new demand value. Whether electromechanical or digital, the electric utility resets the demand value to zero each month after taking its reading for the monthly billing.

Power spikes from in-rush current when starting large motors are averaged in with the power draw from all other equipment operating during the peak demand period. Spikes are typically of such short duration that they have no effect on billing demand regardless of their magnitude.

Although it would be difficult to determine the demand at any given time without a wattmeter, it is not hard to estimate what the maximum demand might be based on knowledge of the power draw of each piece of equipment coupled with the equipment's estimated duration of use (see Chapter 11). In addition, it is important to know the peak demand currently in effect on the power bill and how it would be affected by adding high load units. The electric utility's meter may be equipped with easy-to-read real-time displays of demand and consumption and a power factor that would enable the operator to observe how changes in load would affect demand and consumption.

The *demand charge* is a charge used by the utility to recover its capital or fixed costs of providing power service. The cost for repayment of debt for building power plants, transmission lines, transformers, and rights-of-way are recovered through demand charges. The utility must size its generation capacity to serve the maximum (peak) demand on its system, plus some percentage of gross generation capacity as a safety margin and to sustain load growth. The cost of generating capacity is a fixed cost that, under utility pricing tenets, must be allocated over the customer base on a "just and reasonable" basis. The demand charge assesses each user for fixed charges based on maximum power requirements that have a direct bearing on the size of wires, transformers, and generation capacity provided by the utility for the specific service.

Rates may vary for demands created during predefined peak usage periods versus off-peak or even partial-peak periods. Demand charges may even change seasonally. Whereas most demand charges are created on a billing-month basis, some utilities may include ratchet mechanisms that look at the maximum demand created over the previous 10 to 12 months and assess an "either/or" charge based on whether the current month is greater than, for example, 85% of the maximum demand from the previous 12 months. Often, the demand charge will remain on the customer's bill for the subsequent 12 months after the peak demand occurs.

Another aspect of demand charge is *contract demand.* This is often a minimum charge based on a contract created between the utility and a large customer, such as a WWTP, for providing initial service. This will occur, for example, when a new WWTP is built to service greater future needs. When a WWTP is built in a remote area and requires new electrical service, the electric utility may have to provide new transmission lines and transformers and size them for greater future needs. If initial usage by the facility is low, conventional demand charges will not help the utility meet its debt obligations. The utility may require that the facility owner enter into a contract through which the facility owner agrees to pay a minimum charge in the form of a higher level of demand to obtain the necessary electrical service. This additional charge helps the utility recover its capital investment at a pace more in line with its bond debt repayment.

3.1.4   *Power Factor Adjustment*

The *power factor* is the ratio of actual power and apparent power in an alternating current circuit. *Actual power* is the instantaneous power demand as measured in kilowatts; *apparent power* is the instantaneous power as measured in kilovolt-amperes (kVA), which is the product of the voltage and current being used.

Power factor = actual power/apparent power = kW/kVA

Apparent power being delivered to a WWTP can exceed actual power because of the use of certain common types of devices such as transformers and inductive motors. The difference between apparent power and actual power, which is due to the kilovolt-ampere reactive power (kVAR) is the energy used to produce the induced magnetic forces in transformers and motors. These inductive forces cause a phase shift between voltage and current. (See Figure 2.1 for the relationship of the actual to apparent to reactive power vectors.) The power factor is also defined as the cosine of the phase angle between voltage and current, as follows:

$$\text{Power factor} = \cos(\theta) = kW/(kW^2 + kVAR^2)^{1/2} \qquad (2.1)$$

This equation is useful to calculate the power factor when the utility bill only shows kilowatts and kVAR.

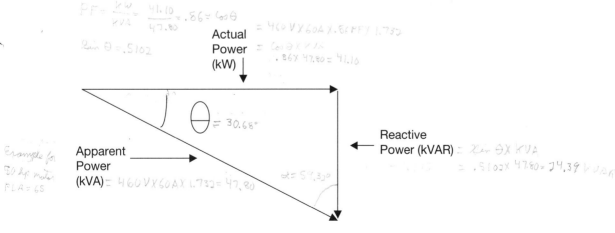

FIGURE 2.1   The power vector relationship.

A problem originating from the power factor is that the size of transmission lines and transformers must be based on apparent power (kilovolt-amperes), but the utility can only recover the costs of actual power in demand and energy charges. Consequently, a low power factor may require the utility to recover capital costs for having to provide oversized equipment to satisfy customers with low overall power factors.

Utilities may use various methods to recover capital costs for a low power factor. A penalty may be assessed if the power factor does not meet a certain standard (e.g., 85%), and credits may be applied if the power factor exceeds 86%. Another less obvious method is to assess demand charges in terms of kilovolt-amperes instead of kilowatts. If the demand charge is assessed in terms of kilovolt-amperes, it is a combined charge for both demand and a low power factor.

The power factor is most commonly corrected by using an automatic capacitor switching bank at the main service, adding capacitors at the motor or starter of large motors, or using synchronous motors. More detail is provided on synchronous motors in Chapter 3.

The power factor is measured using a meter that is separate from the wattmeter and that measures kilovolt-amperes or the power factor directly. A combination meter is also available for this measurement.

3.2 Miscellaneous Charges and Surcharges

3.2.1 Fuel-Cost Adjustment

To compensate for the volatile price of fuels, utilities may have a separate charge for fuel. This allows them to keep published energy charges fairly constant while adjusting for their actual fuel costs on a monthly basis through a fuel-cost adjustment charge.

3.2.2 Regulatory Fees

Regulatory fees are policy- or legislative-adjusted rates that are imposed to satisfy a cost deficiency that results from fuel-cost apportionment or are mandated by legislative or other action.

3.2.3 State and Local Taxes

These taxes can be applied to the billing of both demand and energy as a government revenue source. Most municipal operations are exempt from state and local taxes.

3.2.4 Transmission Voltage

Discounts may be applied to demand, energy, or total amount billed when a user transforms power to lower voltages with its own equipment and at its own expense.

This recognizes the utility's lower capital cost investment. The discount may take the form of a single percentage discount or may vary with service voltage.

3.2.5   Standby Service

Many municipal WWTPs are directly connected to two separate sources of power for greater reliability. The second source is typically only used when the primary source fails. Standby service charges apply to the secondary source, even if not used.

Additionally, many cogeneration/independent power producers use the utility grid as a secondary source of reliability for must-run process electric loads. Some electric utilities charge for the ability to have the electric grid serve as a standby supplier of electric service in the event of forced or scheduled outage of the customer's generation equipment. These rates have the appearance of demand charges applicable to the loads that can be served by the utility.

3.2.6   Nonfirm Power Supply

This concept would provide a price incentive (favorable tariff) to WWTPs whose power supplies could be interrupted or reduced on short notice because the plant has on-site power generation facilities. Nonfirm rates, or interruptible rates, allow the electric utility the choice of diverting power from one customer or customer group to another.

3.3   Other Rate Structures

3.3.1   Flat Demand Rate

Where demand is assumed to be known and fairly constant and metering is deemed unnecessary or not cost effective, a utility may establish a flat demand rate based on a price per kilowatt or horsepower for a certain duration of time. An example of this is external lighting such as street lights, which are often billed on the basis of a flat charge per light per month, variable with the size of the light and with the estimated number of dark hours per month (by season). For example, a rate for a 100-cd (candela) (1000-lm [lumen]) street light may be expressed as a flat charge of $50 per month, subject to a dark hours index, which adjusts the charge per month in accordance with the number of dark hours either measured or estimated by the utility or some other authoritative source.

3.3.2   Flat Energy Rate

The use of flat energy rates can be designed to measure direct variable costs of electric service or other production costs. The flat rate results in a level charge per kilowatt-hour of energy consumption, regardless of the consumption amount.

3.4 Incentives to "Load Shift" and "Peak Shave"

The potential benefits of daily load shifting and peak shaving can be shown using a primary electric service rate schedule of an electric utility in the southeastern United States (Table 2.1). This rate structure is an example of time- and season-differentiated pricing, which provides an incentive to electric utility customers to moderate their energy consumption during peak periods. The rate schedule is as follows:

- On-Peak Winter: Mondays through Fridays during the hours from 7 a.m. to 10 a.m. and 6 p.m. to 9 p.m., excluding Thanksgiving Day, Christmas Day, and New Year's Day;
- On-Peak Summer: Mondays through Fridays during the hours from 1 p.m. to 6 p.m., excluding Memorial Day, Independence Day, and Labor Day;
- Shoulder Period Winter: Mondays through Fridays during the hours from 10 a.m. to 6 p.m., excluding Thanksgiving Day, Christmas Day, and New Year's Day;
- Shoulder Period Summer: Mondays through Fridays during the hours from 11 a.m. to 1 p.m. and 6 p.m. to 8 p.m., excluding Memorial Day, Independence Day, and Labor Day; and
- Off-Peak Period: All hours of the year not covered by the On-Peak and Shoulder Periods.

This rate schedule provides an incentive to users who have the capability of shifting as much of their loads as possible from on-peak and shoulder periods to off-peak periods and/or reducing the consumption charge by operation of peak-shaving on-site power generation facilities, particularly during summer on-peak periods.

3.5 Electric Service Options

The various service options available will depend on the local utility's circumstance. An example of a nonstandard power purchase arrangement offered by utilities is the

TABLE 2.1 Primary electric service rate structure

Time-of-Day Period	Winter consumption rate $/kWh	Summer consumption rate $/kWh
On-Peak	0.05454	0.12351
Shoulder	0.04046	0.08839
Off-Peak	0.0359	0.0577

Winter: November 1 through March 31
Summer: April 1 through October 31

interruptible service contract. As discussed previously, utilities sometimes offer lower rates to industrial customers that have the ability to manage their energy consumption.

Interruptible service means that the customer may be required to cease all use of interruptible electrical service when requested to do so by the utility on short notice. Automatic interruptions may be required, and are controlled by underfrequency relay equipment.

A typical contract may include provisions that the utility may interrupt power deliveries to the customer no more than "x" times per month or per year, with a specified minimum notification period (e.g., 30 minutes to 1 hour). This contract arrangement provides a discount to the customer in return for giving the utility more flexibility in managing its power deliveries, thus optimizing the utility's use of generation capacity. An on-site power generation facility will be needed at the WWTP.

Curtailable service means that on no more than a specified number of occasions the WWTP may be required to cut back a designated amount of electrical demand (measured in kilowatts) when requested by the utility.

3.6 Alternative Energy Sources

Programs are becoming available that allow for the direct purchase of power from alternative "green" energy sources, such as wind and solar power, under long-term contracts with fixed rates. As the cost of power from conventional sources continues to increase, these fixed-rate contracts from green sources will eventually produce savings and reduce the WWTP's carbon footprint. Biogas (methane) produced by the anaerobic sludge digestion process at WWTPs is also an example of an alternative green energy source when it is used in the production of heat for buildings and anaerobic digesters as well as the generation of electricity. In addition to reducing the WWTP's carbon footprint and reducing the cost of electricity, this green power source, when connected to the utility power grid, could also be purchased by the utility at a higher rate as part of the utility's green power program.

3.7 Cogeneration Facilities

A variety of contracts may be required of customers who operate cogeneration facilities. Examples of special contracts include:

- Paralleling agreement,
- Standby power agreement, and
- Standard operating agreement.

A paralleling agreement conveys authority to the cogenerator to operate its cogeneration unit in parallel with the utility power system. *Parallel operation* means operating while connected to the power grid for augmenting power requirements to greater than the generator's abilities in order to supplement or partially offset the WWTP power demand and consumption. In this contract, the interest of the utility is to protect its network and customers from disturbances that could occur from a cogenerator's output. The contract conveys to the cogenerator the right to operate in parallel, while allowing the utility the right to place various demands on the cogenerator. Some utilities do not favor paralleling because they fear it may have an adverse effect on their grid systems.

A standby power agreement compensates the utility for holding a portion of its generation capacity available to serve WWTP loads should the cogeneration unit fail or be taken off-line. A standard operating agreement spells out many of the detailed operating parameters that help maintain the integrity of the utility network. The agreement may also include rules regarding how power plant operations must maintain control and how users should pay for protection equipment that interconnects the two systems.

4.0 SAMPLE ELECTRIC BILLS

Tables 2.2, 2.3, 2.4, and 2.5 provide examples of electric utility bills from 2007 for four WWTPs located in different regions of the United States.

As these sample electric utility bills demonstrate, whereas the rates and structure of electric utility bills vary considerably across the United States, the basic components of demand, energy, and customer charge are similar.

TABLE 2.2 Sample electric bill with demand and peak charges (Southeastern United States)

Charge Item	Use	Units	Rate ($)	Charge
Customer				$23.75
Demand	3146.4	kW	5.70	$17,934.48
Consumption	324,000	kWh (on-peak period)	0.05454	$17,670.9
	439,200	kWh (shoulder period)	0.04046	$17,770.04
	1,123,200	kWh (off-peak)	0.0359	$40,367.80
Total				$93,767.03
Average cost per kWh ($)				0.049

TABLE 2.3   Sample electric bill with power factor credit (Midwestern United States)

Charge Item	Use	Units	Rate ($)	Charge
Consumption	3,460,800	kWh	0.02769	$95,832.21
Demand	6307	kW	10.91	$68,824.99
Power Factor	97%			$(5,759.01)
Total				$158,898.19
Average Cost per kWh ($)				0.0459

TABLE 2.4   Sample electric bill with consumption and demand peak period charges (Northeastern United States)

Charge Item	Use	Units	Rate ($)	Charge
Consumption	140,800	kWh (on-peak)	0.0446	$6,279.68
	121,600	kWh (shoulder)	0.0356	$4,328.96
	104,000	kWh (off-peak)	0.0218	$2,267.20
	366,400	kWh	0.0938	$34,371.62
Demand	651.2	kWh (on-peak)	12.36	$8,051.44
	1276.8	kWh (shoulder)	4.416	$2,659.76
Miscellaneous				$1,199.91
Sales Tax			0.0875	$5,171.96
Total				$64,280.11
Average Cost per kWh ($)				0.175

TABLE 2.5   Sample electric bill with wind power credit (Northwestern United States)

Charge Item	Use	Units	Rate ($)	Total
Energy	4,881,600	kWh	0.046955	$229,215.53
Demand	13,480	kVA	3.14	$42,327.20
Reactive Power	2,736,000	kVARh	0.00*	$0.00
Electric Conservation Program	4,881,600	kWh	0.00105	$5,125.68
Power Cost Adjustment	4,881,600	kWh	0.00	$0.00
Wind Power Production	4,881,600	kWh	(0.000959)	($4,681.45)
City Tax			0.067	$18,223.13
Total				$290,210.09
Average Cost per kWh ($)				0.059

* Using the values listed for kWh and kVARh in Equation 2.2, the average power factor calculates to 87%.

5.0　NATURAL GAS BILLING

The basic rate components of demand, energy, and customer charge are similar in concept for those of natural gas. In addition, the same variations on rates, based on peak day and season, are typically applied.

With natural gas, energy is measured by units of gas consumed, measurable in hundreds of cubic feet (ccf) (cu ft $\times$ 0.02832 = m^3), thermal heating value (therms), or millions of British thermal units (MMBtu) (1.055 GJ):

- 1 therm = 100 000 Btus,

- 1 MBtu = 10 therms, and

- 1000 cu ft (MCF) equals approximately 1 000 000 Btu or 1 dekatherm.

Because 0.03 m^3 (1 cu ft) of methane has a heating value of approximately 1055 kJ (1000 Btu), 3 m^3 (1 ccf) of natural gas has a heating value of approximately 1 therm.

Recently, the trend has been to offer separate rates for the commodity itself (therms of natural gas) and its associated pipeline transportation charges and for local distribution of the commodity through the local utility's pipelines to the customer. These changes in rate structure are in response to deregulation and to a growing number of independent gas producers that offer natural gas at discounted prices, but whose existing distribution lines are owned and operated by established utilities to deliver the commodity to the customer. In addition, there could be other charges such as balancing, storage fuel charges (for operating compressors), and losses in the pipeline transportation from the hub.

5.1　Rate Structures

Billing determinants for gas service vary from state to state and also depend on the industry arrangement present in the local market. The following is a general view of the diversity of industry status.

5.1.1　Unbundled Utility Service

Through deregulation, competitive markets have developed in some states as alternatives to the traditional monopolistic markets of local gas suppliers. A "noncore" distinction is made for customers that have alternative choice(s) for procuring gas or other economic fuel alternatives. In the noncore market, gas services have been disaggregated so that gas procurement, gas transportation, delivery balancing, and storage services can be separately procured and charged.

5.1.2   Transportation

Some states distinguish gas services between traditional procurement-transportation-supply service and transportation service only.

5.1.3   Pipeline Direct

A third condition exists in some areas where Federal Energy Regulatory Commission-regulated interstate pipelines service industrial demands with a direct connection from a transmission line and bypass the local distribution company serving the surrounding area.

The billing components of gas service and transportation service are similar to those of electric utilities as discussed previously:

- *Demand* is the contracted level of service indicated by the customer's experienced peak usage on the system's peak day or a negotiated maximum until operating experience justifies a change. Demand may also be assessed using monthly peak or moving average peaks to assess the system service rendered.

- *Demand charge* is a monthly fixed charge that is applied to the measured or aforementioned contract demand.

- *Energy,* or *throughput,* is a variable rate assessed on the actual gas throughput that has been conveyed or delivered.

- *Interruptible service* in the gas industry is typically limited to large customers. Interruptions are often based on community heating load conditions. The level of interruptibility is a negotiable element in service rates.

5.2   Rates

Rate categories vary for each utility with regard to terminology and the conditions that apply to each class of service. General service rates apply to small users, are typically at a firm rate, and are noncurtailable in many parts of the country. Industrial rates vary more seasonally, may require a minimum usage per month year round, and vary in curtailability with the utility. Large users will find industrial rates attractive for cost savings, particularly in summer months, but the frequency of curtailability must be tolerable by either reducing usage or switching to a backup fuel. Additionally, unbundling of natural gas rates in some states has resulted in the availability of special contracts for natural gas transportation and supply.

5.2.1   Gas Transportation Agreement

The gas transportation agreement covers transportation services the utility provides in moving gas from the state border (or in-state gas fields) to the customer delivery point. The contract may provide that the customer will be served under a cogeneration transportation rate schedule, which provides that the transport cost must be no more than the equivalent service to the utility's gas boilers that generate electricity. The rate structure has components that vary with the volume of gas used and measured monthly and other components that fluctuate on a rolling basis of 12 months, measured on the highest demand at any single point in time.

Gas companies typically offer transportation programs that may be attractive to large year-round users. These are similar to commodity contracts in that the user agrees to buy a specific quantity in a specific time period at a specific price. Cost savings for a large user can be significant; however, the risk of committing to price, quantity, and delivery time must be carefully weighed.

The transportation of gas from well field to the local distribution company can be performed by the purchaser (in the case of large industrial loads) or by a gas broker. Once the transportation route and user's operating needs are addressed, the local gas company will assist with ultimate delivery.

The local gas company agrees to arrange transportation from the pipeline interconnect to the user typically for a fixed fee per million British thermal unit (MBtu) and a fixed monthly administrative charge. The gas company may require that the customer commit to a specific quantity of a gas delivered for the following month by day 5 of each month. Gas purchased under this type of program is typically curtailable to 0%. In addition, there will typically be a minimum quantity per month to be on the program, and gas quantities used beyond the contract amount will be charged at an industrial rate.

To illustrate how supply and transportation of natural gas can be billed, Table 2.6 presents an example of a separate natural gas billing for the supply of natural gas to the city gate, and Table 2.7 presents an example of a separate natural gas billing for the transportation from the city gate to an Upper Midwest WWTP in 2008.

5.3   Seasonal Pricing Incentive

Prices for natural gas vary with the market conditions for supply and demand. Because demand is high in winter, prices are typically highest then and lowest in the summer months. Unfortunately, WWTPs follow this pattern of high demand in

TABLE 2.6   Example of natural gas billing for supply

Charge Item	Use	Units	Rate ($)	Charge
Supply	15,886.45	Dekatherm	9.44	$149,968.09
Balancing and Nominating	15,886.45	Dekatherm	0.05	$794.32
Total				$150,762.41

winter because space heating needs and process demands such as sludge heating peak in colder months. Some WWTPs may have peak electrical power requirements in summer and may try to take advantage of low summer natural gas prices (or digester gas produced on site) to fuel engines that drive generators, pumps, or blowers. If this method is used with natural gas to *peak shave* or provide *prime* power, a more constant pattern of usage may be established over the entire year to take advantage of off-peak rates available from gas companies. A WWTP with an anaerobic digestion system that produces digester gas may need natural gas for only the coldest months; thus, it will be difficult to get anything but natural gas at the highest rates for the limited time of usage. In any case, it would be a good idea to look at the WWTP's gas usage patterns and discuss them with the gas company representative at least every 5 years to make sure that the WWTP is getting the best rate available to meets its needs.

5.4   Computing Thermal Consumption

Calculating the therms used requires a knowledge of the volume of gas used as well as the heating value. For natural gas, the heating value is available from the gas company and will normally range from 37 000 to 40 000 kJ/m^3 (1000 to 1080 Btu/cu ft).

TABLE 2.7   Example of natural gas billing for transportation

Charge Item	Use	Units	Rate ($)	Charge
Customer Charge	Large			$673.17
Distribution Margin	158,864.50	Therms	0.0343	$5449.05
Demand	8110.00	Therms	0.1475	$1196.23
Total				$7,318.45

* *Note:* In this example, the total average cost of supply and transportation of natural gas was $0.995 per therm.

Liquid petroleum gas heating values are also available from the gas supplier and are typically 26×10^6 kJ/m^3 (92 000 Btu/gal).

Because gas is a compressible fluid, it is not always sufficient to know the volume of gas without knowing its pressure. Natural gas is supplied at varying pressures, and the compressed volume delivered must be converted to a volume at standard conditions of temperature and pressure. It should also be noted that the moisture content and hydrogen sulfide concentration are important in determining the corrosiveness of the gas.

For natural gas customers with an interruptible service, a backup fuel is required. For equipment such as burners for boilers, a dual fuel burner will serve this purpose. For smaller equipment such as space heaters, different burner orifices are required to fire liquid petroleum gas and different burners (and heaters) for fuel oil. Therefore, it is important to have a portion of the noninterruptible service to serve those smaller units that cannot be operated with dual fuels.

6.0 UTILITY RATE AND SERVICE OPTIONS

The best way to find out about the types of electric and gas rate structures available is to contact the local utility provider and request:

- Tariff schedules are available for industrial customers and the special requirements to qualify for one rate schedule versus another.
- Copies of the tariff pages and sample contracts for review and analysis.

In some cases, lower rates may be available to customers with high average loads. In other cases, special contracts may be available to customers who have some flexibility in their energy consumption patterns. For example, customers who have the ability to interrupt or curtail energy deliveries may qualify for special rates. In preparing for initial discussions with utility company representatives, WWTP loads—both energy and demand—for 12 consecutive months should be studied and graphed to fully understand the potential benefits of the various rate and service options offered.

Chapter 3

Electric Motors and Transformers

(continued)

1.0 THE NEED FOR EFFICIENT MOTOR DESIGN

Water Environment Federation (WEF) has estimated that 90% of the electrical energy consumed in a typical wastewater treatment plant is used in the operation of electrical motors. The increasing cost of electrical power has driven demand for more efficient motors and electrical equipment. Numerous U.S. and international standards organizations have addressed this issue through the revision of existing standards or the development of new standards. The U.S. government, through the U.S. Environmental Protection Agency (U.S. EPA), the U.S. Department of Energy (DOE), and other agencies, has issued many handbooks and guides that promote more efficient electrical designs.

Illustrating the cost of efficiency, U.S. EPA has estimated that during a 10-year operating life a motor can consume 50 times its initial purchase price. This means that a $2,000 motor, if operated continually over 10 years, can translate to more than $100,000 in electrical costs. Based on these facts and the fact that energy costs are steadily increasing, the owner and engineer should strongly consider life cycle costs before the initial cost.

A second issue facing owners and engineers is that the initial design requirements address maximum expected load. In reality, this load level may not be reached until sometime in the distant future, if at all. Consideration should be given to staging installation of equipment over a period of time to allow for the better matching of operating loads.

There are numerous types of motors used in wastewater and water plants, from small, fractional horsepower (hp) motors to large motors. This chapter will focus on those motors that are significant contributors to energy consumption.

1.1 Motors as Part of a System

Development of the microprocessor and movement of this device from the control room to the end device have radically changed the overall design of the motor. A motor package today consists of multiple components joined seamlessly together. The smart motor may combine a microprocessor for communications, a controller for loop functionality, imbedded sensors for monitoring motor health, and diagnostic software for operations and performance. *Mechatronics* is a new term used to describe this combination of previously separate functions into one package. The boundary line between mechanical, electrical, electronic components, and software is now blurred. The technician who works on today's motors must have an understanding of all of these areas. This chapter will focus on the mechanical and electrical aspects of the design.

1.2 Components of Common Motors

The design of the electrical motor has been around for many years. Every motor consists of a few basic components (see Figures 3.1, 3.2, and 3.3). Improvements in the

FIGURE 3.1 Motors (courtesy of Baldor Electric Company).

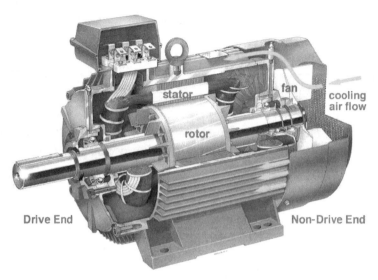

FIGURE 3.2   Alternating current—motor cut-away view.

1. Bearings

2. End Cap—Shaft End

3. Brushes

4. Insulated Windings

5. Brush Holder

6. Commutator

7. Frame

8. Armature/Rotor

9. End Cap—Base

FIGURE 3.3   Direct current—motor (with gear) cut-away view (courtesy of LEESON Electric—A Regal-Beloit Company).

design of each of these components and the optimization of their alignment to one another has led to a new generation of high-efficiency motors.

The armature, or rotor, is the part of the motor that is made to rotate by the changing magnetic field. The stator, or field, is the stationary component that causes the magnetic field to change. The shaft, connected to the rotor, provides the primary connection for the conversion of electrical/magnetic energy to rotational force. The "air gap" is the clearance distance between the rotor and stator. Bearings are designed to minimize the frictional forces of the shaft's rotation. The movement toward improving the energy efficiency of the electric motor has meant that the design and manufacture of each of these components has to be optimized. This topic is further explored in section 7.0 of this chapter ("Standards for Energy-Efficient Motors").

2.0 MEASUREMENT OF ELECTRICAL CHARACTERISTICS

The characteristics of the power supplied to a motor and the behavior of the electricity within the motor can be measured. These measurements are used to assess the quality of the power supply, the load being placed on the motor, the motor's general operating efficiency, and the motor's electrical integrity. It is important to note that all electrical tests and measurements must be performed by personnel specifically trained for this technology.

2.1 Voltage

For diagnostic purposes, three-phase voltage measurements are taken for the three leads (L_1, L_2, and L_3) on the load side of the motor controls (Figure 3.4). The voltage unbalance for the measurements obtained between L_1 and L_2, L_1 and L_3, and L_2 and L_3 should be less than 1%.

$$\text{Unbalance (\%)} = \frac{\text{Maximum deviation average}}{\text{Average of the three readings}} \times 100 \qquad (3.1)$$

If the unbalance exceeds 1%, the amperage unbalance and the motor operating temperature will be high. Frequently, problems with the motor controls or power supply result in voltage unbalances. Testing for voltage unbalance should be part of equipment preventive maintenance.

Motors are designed to operate within a narrow voltage range. Typically, the acceptable voltage range for a motor is the nameplate rating ±10%. Voltage tolerance levels are defined by ANSI/IEEE C84.1 (NEMA, 2006), a joint standard of the American National Standards Institute (ANSI) and the Institute of Electrical and Electronics Engineers (IEEE). The manufacturer's literature for a specific motor should be consulted to ensure proper operation.

2.2 Amperage

The currents (amperage) flowing through each phase (L_1, L_2, and L_3) of a three-phase motor must be close to equal. When the amperage between the phases is unbalanced, the motor runs hot and loses efficiency. Amperage measurements should be taken when the motor is at its normal operating temperature and loading. The percent unbalance is calculated using the same formula used for voltage unbalance. At full load, the maximum allowable unbalance is 5%. If a motor is operating at considerably less than full load, a larger percentage of unbalance may be acceptable.

Identifying the source of voltage or amperage unbalance requires some troubleshooting. (Again, it is important to note that all electrical troubleshooting must

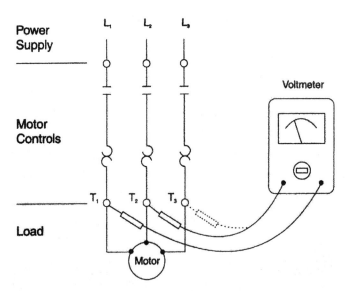

FIGURE 3.4 Checking three-phase motor voltage (L = power lead and T = motor starter terminal). (Reprinted with permission from ITT FLYGT Corp. [1985] *FLYGT Product Education Manual*. 8th Ed., Product Education, Norwalk, Conn.)

only be performed by qualified personnel who have been trained in this technology.) First, note the voltage and amperage with the original arrangement of power leads. Next, change the location of all three power leads on the load side of the motor controls and repeat the measurements. It is important to make sure that the positions of all three leads are changed. If only two leads are changed, the motor will operate in the reverse direction. Change all three leads to produce the third possible arrangement and take a third set of readings. Figure 3.5 provides an example in which the black lead always has an amperage of 19. This indicates that the problem is in the motor or the wiring leading to it. If the same lead always produces the high reading, the problem is somewhere between that terminal and the power plant. The problem could be as simple as worn starter contacts or a loose terminal screw. Conversely, assistance may be required from the power company for a solution.

2.3 Power Factor

The easiest and most accurate method available to measure the power factor is to use a power factor meter. The power factor is the ratio of the actual power (usually expressed in watts) to the apparent power (usually expressed in voltamperes). This relationship is illustrated by the power triangle shown in Figure 3.6.

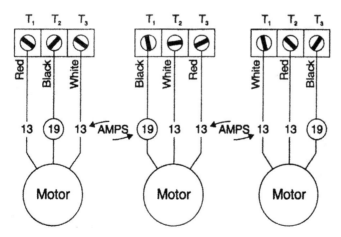

FIGURE 3.5   Unbalance caused by load (T = motor starter terminal). (Reprinted with permission from ITT FLYGT Corp. [1985] *FLYGT Product Education Manual*. 8th Ed., Product Education, Norwalk, Conn.)

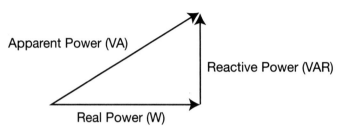

FIGURE 3.6 Motor triangle.

The kilovolt-ampere reactive power (kVAR) is the energy used to produce the induced magnetic forces that make the motor rotate. The reactive power is not measurable with a wattmeter. The power factor can also be obtained by calculation, using data obtained with a wattmeter, voltmeter, and ammeter. The real power (kilowatts) and apparent power (kilovolt-amperes) are measured directly and used in the calculation. It should be noted that power costs are based on real (kilowatts) and not apparent power.

2.4 Resistance and Insulation

The flow of electricity from a point of high potential to a point of low potential is governed by the resistance of the space between the two points. An electrical conductor provides a path of low resistance to facilitate the flow of electricity along an intended path with minimal losses. Insulation confines the flow of electricity within its intended path by shielding the electrical path along a conductor with a barrier of high resistance. Over time, the insulation around conductors begins to break down and allows current losses to ground. Insulation on wires leading to the motor and on the motor coils themselves should be checked routinely as part of preventive maintenance. This simple measurement using a megohmmeter ("megger") can help prevent emergency maintenance situations. (Before any testing, it is important to verify that all safety procedures are being observed and security is in place.) Because the resistance of the insulation to the flow of electrons is being measured, the resistance values are measured in megohm (million ohm) units. Whenever resistance is being measured, all sources of power must be removed from the circuit being tested. It is important to always check for voltage before putting the megohmmeter probes on the terminals. The megohmmeter has its own power supply to energize the circuit being tested. Each power lead to the motor is checked as shown in Figure 3.7. A new motor should read infinite resistance. Over time, the resistance will gradually

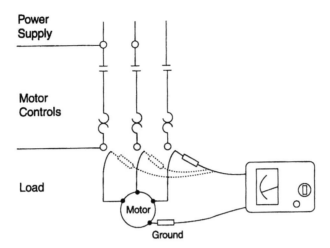

Power
Supply

Motor
Controls

Load

Motor

Ground

FIGURE 3.7   Checking motor insulation, coil to ground, for a megohmmeter. (Reprinted with permission from ITT FLYGT Corp. [1985] *FLYGT Product Education Manual.* 8th Ed., Product Education, Norwalk, Conn.)

decrease. This information can be graphed so that the trend can be visualized. When a minimum resistance value is reached, the coils can be dipped and rebaked to restore the insulation value. Some manufacturers allow 2 MΩ/hp as the minimum resistance value before the coils are serviced. For small motors, a somewhat higher ratio should be used. If the motor is checked at the motor controls, this activity is also checking the insulation of the conductor to the motor. Before the motor coils are faulted for low resistance, additional readings must be made at the motor. Large motors should consider use of an online megger.

2.5 Power

Measured power is the actual instantaneous electrical demand required to operate a motor under a full or partial load condition. Power is measured in kilowatts and may be obtained directly with a wattmeter. Knowing the power consumed by a motor can be useful in evaluating the operating efficiency of the system. Power measurements can be compared to the motor's power rating to determine whether the motor is overloaded (undersized) or underloaded (oversized). Overloading may be the result of poor motor selection or of a motor, drive, or driven element problem, which requires immediate attention. Motors are most efficient at their full-load condition.

2.6 Slip

Slip is the difference in actual rotor speed and the synchronous speed of the motor, as follows:

$$\text{Slip} = N_S - N \tag{3.2}$$

Where

   N_S = synchronous speed, r/min (revolutions/minute), and
   N = actual speed, r/min.

Load estimates can be made from motor-operating characteristics such as current, slip, and input power. Figure 3.8 shows typical performance curves for a 7.5-kW (10-hp) squirrel cage motor. As can be seen, current is directly proportional to output power in only a narrow range—approximately 70 to 110% of full load. Thus, overestimates of load will be made in cases where the motor is lightly loaded. Slip, how-

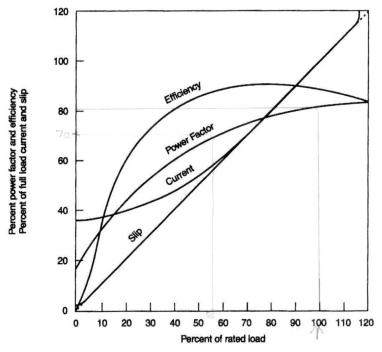

FIGURE 3.8   Typical performance characteristics for a 7.5-kW (10-hp) squirrel cage induction motor as a function of load.

ever, is directly proportional to load throughout the operating range and allows a method of load determination at any operating point. Load can also be calculated from input power, but the efficiency at the operating point must be known. Load determination from slip and input power measurements are discussed in the next paragraph. Slip measurements provide a convenient means of estimating motor loading, provided accurate and precise measurement devices are available.

Because slip is proportional to load, the ratio of the measured slip to the full-load nameplate slip yields the relative motor load. The slip of a squirrel cage induction motor is small (usually less than 3%). A stroboscopic tachometer with an accuracy of 0.1% or better is required to produce meaningful results. An error of only 1% would result in an error of 30% in the slip measurement. Furthermore, considerable error may be inherent in this method when using manufacturers' nameplate data because the National Electrical Manufacturers Association (NEMA) standards allow a 20% deviation in the nameplate slip.

3.0   OPERATING POWER

Motors are rated in units of watts (W) (horsepower) on the basis of the maximum amount of work that they are capable of performing. The actual power consumed at any given time by an operating motor is in direct proportion to the work being performed. In other words, if a 75-kW (100-hp) motor is only doing 60 kW (80 hp) of work, the motor is only supplying the equivalent electrical power of 60 kW (80 hp). In general, motors 10 to 20% larger than actually required are usually selected by the engineer in the design process. It is important to note that the common convention is to use kilowatts in place of watts.

Nameplate power is the output power of the motor; it is also known as the *shaft power.* Input power to the motor is always greater than output power. For example, a 75-kW (100-hp) motor doing 75 kW (100 hp) of work will probably be drawing 80 kW (110 hp) from the power supply. Because motors are commonly oversized by 10% and are only 90% efficient, nameplate power is often taken at face value for power draw in energy calculations.

3.1 Service Factor

A 75-kW (100-hp) motor on a pump that needs to do 110 kW (150 hp) of work will most likely fail. The service factor is a number placed on each motor by its manufacturer indicating the safe amount greater than the nameplate rating at which the motor can run without failing. It is unwise to install a motor in an application with a design rating in excess of the motor nameplate. The service factor allows for above-design operating power requirements caused by such factors as bearing problems. "Service factor" is stamped on each motor nameplate and is expressed as a percentage of the nameplate power rating, for example 1.15 or 115%. This is the most common service factor. Motor service factors should only be used as a safety factor when the load temporarily exceeds the nominal rating of the motor.

4.0   TYPES OF ELECTRIC MOTORS

Motors are available for direct current or alternating current applications, with alternating current motors also being available for single-phase or three-phase current. Motors are also available for many standard voltages. Small motors are typically single-phase and operate on a 110-V electric current. Larger motors found in WWTPs are typically three-phase and operate on the 480 V three-phase power that normally serves industrial applications. Large motors with higher voltage requirements are needed when greater horsepower is required.

All alternating current electric motors operate on the same basic principles regardless of type or size. Rotation of the shaft is the result of the force created by the interaction of a magnetic field and the current between rotor (rotating electrical coil) and stator (stationary electrical coil). It makes no difference whether the magnetic field is created in the rotor or the stator. Within this simple principle, there are many different types of alternating current motors, each having its own operating characteristics specifically suited to the drive application.

4.1   Three-Phase Motors

Because of their high efficiency and low maintenance requirements, three-phase squirrel cage induction motors are the most common type in use. The efficiency of squirrel cage motors is affected by a number of factors, including the characteristics of the electrical power supplied and the motor's size, design, electromechanical integrity, and operating load. The last two factors can be partly controlled by the operator through proper operation and maintenance (O&M) practices. Motorized operations should incorporate the most practical energy-efficient systems available during the design process. Comparing competing systems on the basis of annual worth for capital plus O&M costs will facilitate the proper selection.

There are three general types of three-phase motors that all have unique rotors and operating characteristics. The three types are the squirrel cage induction motor, the wound rotor induction motor, and the synchronous motor. The three-phase electric power required for these motors, regardless of whether they are induction or synchronous motors, can be calculated as follows:

$$P = V \times I \times \sqrt{3} \times PF \tag{3.3}$$

Where
> P = power, W;
> V = line voltage, V;
> I = average line current of 3 legs × square root of the number of phases, A;
>   and
> PF = power factor.

4.1.1 Squirrel Cage

The most common industrial motor is the squirrel cage induction motor. It is the least expensive type of induction motor and is available in all common power ratings and synchronous speeds. Voltage is applied directly to the stator or primary winding. The rotor, or secondary winding, consists of bars of aluminum or copper connected together at both ends by a conducting ring in an arrangement resembling a squirrel cage. The rotor windings form a complete closed circuit with no external connections. The absence of moving electrical contacts makes these motors quite reliable.

The speed of a squirrel cage motor is nearly constant over the normal range of loads. The speed of all induction motors is determined by the line frequency, the number of poles in the motor, and the slip. The synchronous speed of the motor is the speed in synchronism with the frequency of the electric current or the speed at which the magnetic field revolves. It depends on the line frequency and the number of poles built into the motor, and is determined as follows:

$$N_S = \frac{120f}{P} \tag{3.4}$$

Where
> N_S = synchronous speed, r/min;
> f = frequency (60 Hz); and
> P = number of poles per phase.

Synchronous speeds are typically available at 900, 1200, 1800, and 3600 r/min. For a squirrel cage motor to produce torque, the operating speed must be less than the synchronous speed. The difference between the operating speed and the synchronous speed, which is called *slip*, is proportional to the torque produced, and is usually less than 3% of synchronous speed at full load.

Squirrel cage motors are classified by NEMA into four design types (A through D) based on torque, slip, and starting characteristics. Type A has low torque, low slip, and normal starting current. Type B is similar to type A, but draws less current. Type C has more starting torque than types A and B, but cannot bring all loads up to full speed. Type D has high starting torque and high full-load slip.

4.1.2   Wound Rotor

In the wound rotor induction motor, unlike the squirrel cage motor, the free ends of the secondary windings are brought out to slip rings through brushes and are externally connected. By varying the external resistance, the speed can be controlled. Without external resistance, this motor behaves similarly to the squirrel cage induction motor. Energy is expended in the external resistance, and efficiency varies approximately in proportion to the speed reduction. Use of the wound rotor motor is typically reserved for applications in which high starting torques and intermittent operation are required, such as for hoists and cranes.

4.1.3   Synchronous Motors

In a synchronous motor, direct current voltage is applied to the rotor, which locks in step with the rotating magnetic field of the stator causing the rotor to rotate at synchronous speed. However, the synchronous motor, by itself, is incapable of producing torque at other than synchronous speed; thus, alternative starting mechanisms, such as a squirrel cage winding, are required. In the industrial field, synchronous motors are used only in the larger power range, typically 75 kW (100 hp) or greater.

In contrast to the induction motor, a synchronous motor's power factor is not a function of load or size, but is under the control of the user. By adjusting the direct current, the power factor can be changed to a leading, lagging, or unity value. Thus, synchronous motors may be used to offset the power factors of inductive loads.

4.2   Single-Phase Induction Motors

Single-phase current is not typically produced in the United States, but it is readily derived from any two lines of a three-phase supply system. Because of the higher costs associated with local transmission and voltage step-down of three-phase current, single-phase, low-voltage current is supplied to homes and farms. Unlike three-phase motors, single-phase motors are incapable of self-starting and require an auxiliary winding. Motors for single-phase current are more complex than three-phase motors because they require additional devices such as switches and capacitors to induce the rotating magnetic field. Consequently, single-phase motors are generally more expensive, less efficient, less reliable, and have shorter service lives than three-phase motors. For many small-power applications, however, the extra motor cost is outweighed by the costs of additional wiring and motor controls. Additionally, low-power, single-phase motors feature the added convenience of being able to be plugged into the nearest outlet. There are many types of single-phase motors, including shaded-coil, inductive-split-phase, capacitor, repulsion-start/induction-run, and repulsion-induction. For industrial applications, capacitor motors are typically used. The most common capacitor motors are the capacitor-start motor, the capacitor-start/capacitor-run motor, and the permanent-split-capacitor motor.

Although the single-phase motor may not be the best fit for most applications, it has been found that some smaller remote sites may only have single-phase motors available. Using single-phase motors will typically result in higher energy costs and possibly shorter motor life; however, in these special cases there may be no option. Designers should strive to use the highest quality motors in these situations to minimize maintenance issues. Small fraction horsepower applications may also be a good fit for single-phase motors where availability is a factor. These motors have low current draw and have minimum effect on overall energy usage. An alternate approach to the use of three-phase motors with single-phase power availability is to use a phase converter and/or variable frequency drive (VFD). Although not as efficient, this approach will provide a mechanism for a remote application.

4.3   Direct Current Motors

An advantage of direct current motors is precision control and, because of this, they are often used as controllers in automation applications. They are also used for other applications requiring precision, such as cranes, hoists, and elevators. Additionally, they are used in battery-powered mobile equipment and for many electric railway operations. In WWTPs, large direct current motors have been used in variable-speed control applications such as influent and return activated sludge pumping using rheostats or variable-voltage controllers. Because it is impractical for large power producers to generate and distribute direct current voltage, voltage for direct current motors is typically derived near the application from alternating current through various types of rectifiers.

5.0   CONVENTIONS FOR SPECIFYING MOTOR PERFORMANCE

5.1   Definition of Efficiency

Motor efficiency is the ratio of motor output to motor input. It is a measure of how well a motor converts electrical energy to mechanical energy and is usually expressed as a percentage.

$$\text{Motor efficiency} = \frac{P_m}{P_e} \times 100 \tag{3.5}$$

and

$$P_0 = P_e - P_L \tag{3.6}$$

Where

P_m = mechanical power output, W;
P_e = electrical power input, W; and
P_L = power loss, W.

Power loss is that part of the input power that is converted to heat rather than useful work. This loss can be attributed to friction and windage, electrical resistance losses in the rotor and stator, magnetic core losses (hysteresis and eddy currents), and stray load losses.

5.2   Test Procedures

The Institute of Electrical and Electronics Engineers standard IEEE-112 (IEEE, 2004) is a recognized test procedure in the United States that is used to determine motor efficiency. This is further covered later in this chapter. Caution should be exercised when comparing efficiencies based on U.S. standards with those of foreign manufacturers as there are differences in international testing methods that result in different levels of premium motors.

6.0   MATCHING MOTORS TO LOAD

The primary goal in selecting a motor is to be sure the motor selected will have adequate power to drive the intended load throughout the entire anticipated load range. A secondary goal is that the motor be energy efficient and economical.

Motor efficiency for standard induction-type motors increases with increasing motor size and is relatively constant for a given motor at various loads. However, most motors are slightly more efficient at full load, with efficiency dropping off slightly as load decreases from full to half load. For motors of greater than 0.75 kW (1 hp), the change in efficiency from full to half load is generally less than 5% of the full-load efficiency. Similarly, a power factor for a standard induction motor increases with motor size, but the power factor drops off more significantly with a reduction in load. The decrease in power factor for a 0.75-kW (1-hp) motor at half load is approximately 25% of the power factor at full load.

Power factors decrease 10 to 15% or more as load is reduced from 100 to 50%, depending on the power and revolutions per minute of a given motor. Because electrical utilities often assess charges for a poor power factor, it is advantageous to closely match motor size to load to maintain a high power factor at a WWTP. The relationship of both efficiency and power factor to changes in motor load differ between the motor enclosure (open drip-proof, totally enclosed fan-cooled, and others), number of revolutions per minute, and manufacturer.

When studying equipment, if a significant mismatch in the load and electric motor rating is found, it may be economical to replace the motor in question with a motor that more closely approximates the required load. This, however, may not be true for all situations. For example, a 15-kW (20-hp), 1200 r/min, high-efficiency motor is driving a pump and found to be operating at approximately half load. Pump tests verify that a 7.5-kW (10-hp) motor would be able to operate throughout the operating range of the pump as installed without being overloaded. It is assumed that the motor operates at a constant load of 7.5-kW (10-hp). An estimate of savings would be obtained by replacing the 15-kW (20-hp) motor with a 7.5-kW (10-hp) motor of the same type.

7.0 STANDARDS FOR ENERGY-EFFICIENT MOTORS

As defined previously, *motor efficiency* is the ratio of mechanical power output to the electrical power input, usually expressed as a percentage. Numerous U.S. and international standards now define different versions of high-efficiency motors. This section will focus on the primary standards used in the United States.

High-efficiency motors differ from standard-duty motors in terms of higher quality materials, more precise machining and manufacturing, more precise control over air gaps and tolerances, better bearings, more copper in windings, and extended length. This premium cost typically allows for better warranty protection from most manufacturers. Because of these improvements, energy-efficient motors have higher service factors, longer insulation and bearing life, lower waste heat output, and less vibration.

The following section cites agencies that are currently involved with energy efficiency for electric motors and their key programs. However, energy efficiency is a rapidly developing area of interest and new information is continually being added. As such, the reader should check with the agencies for the most current information. There are also numerous international standards that cover this topic. Therefore, a complete listing of all of the agencies involved with energy efficiency for electric motors is beyond the scope of this manual.

7.1   Energy Policy Act of 1992

Because motor operation constitutes a substantial percentage of overall energy consumption, Congress granted the U.S. Department of Energy (DOE) the authority to set minimum efficiency standards for certain classes of electric motors. Congress enacted the Energy Policy Act of 1992 (EPAct), which focused on reducing the total amount of energy used in the United States. The Energy Policy Act's rules for motors became effective on October 24, 1997. However, these rules did not immediately bring about new motor designs. The Energy Policy Act also granted large exclusions for existing motors. (Visit http://www.epa.gov/radiation/yucca/enpa92.html for more information on EPAct.)

7.2   Department of Energy—Energy Efficiency and
Renewable Energy Program

The U.S. DOE developed the Energy Efficiency and Renewable Energy Program to provide linkage between the National Energy Policy and specific energy goals. This program addresses both energy efficiency and alternate energy sources. This is an active and growing program that has numerous test and education programs aimed at reducing national energy consumption. In addition, U.S. DOE's Motor Challenge Program is an industry/government partnership designed to help industry reduce energy consumption during specific time intervals (i.e., duringsummer usage).

7.3   U.S. Environmental Protection Agency

U.S. EPA has moved steadily in developing different programs and documents that support the use of energy-efficient motors. Several of these papers are directly related to the water and wastewater industry. The U.S. EPA Energy Star Program estimates that about $4 billion is spent annually on energy costs to run drinking water and wastewater utilities. They are seeking to reduce this cost through education and the development of more efficient equipment.

7.4   Independent Energy Companies

There are several independent companies that have been established as nonprofit corporations that focus on energy efficiency. One such company, Advanced Energy (Raleigh, North Carolina), is an example of an independent nonprofit corporation that is governed by a board of directors appointed by the governor of North Carolina and member utilities. Numerous motor-related programs are available through Advanced Energy.

7.5   International Motor Standards

As in the United States, the international market has also addressed energy efficiency. This has resulted in many new international standards as well as revisions of old standards. However, it is beyond the scope of this manual to provide a reasonable review of all of the current international motor standards that affect design efficiency. It should be noted that several of these standards have different methods of determining their "premium" motor that are different from the "NEMA Premium" motor (see section 7.6.1. for a description of the NEMA Premium motor program). As such, the engineer should closely examine these differences to determine if they still meet their design criteria.

7.6   National Electrical Manufacturers Association

7.6.1   *National Electrical Manufacturers Association Premium Motors Program*

Although the minimum standards set by DOE were reasonable at the time they were drafted, industry noted the need for more efficient motors. In 2001, NEMA established standard MG-1, which defined energy-efficient "premium" electric motors; the standard was later updated in 2006 (NEMA and ANSI, 2006). These standards were more specific and more stringent than those required by EPAct motors. The NEMA Premium efficiency electric motor program defines the parameters products must meet or exceed to be considered a NEMA Premium electric motor. The NEMA Premium efficiency levels are contained in Tables 12-12 and 12-13 in NEMA Standards Publication MG-1-2006. In this manual, Tables 3.1 and 3.2 list the national efficiencies for NEMA Premium induction motors.

TABLE 3.1 Nominal efficiencies for "NEMA Premium®" induction motors rated 600 V or less (random wound) (Tables reprinted by permission of the National Electrical Manufacturers Association, Rosslyn, VA. NEMA MG1-2006. NEMA Premium®)

	OPEN MOTORS					
	2 POLE		4 POLE		6 POLE	
HP	Nominal Efficiency	Minimum Efficiency	Nominal Efficiency	Minimum Efficiency	Nominal Efficiency	Minimum Efficiency
1	77.0	74.0	85.5	82.5	82.5	80.0
1.5	84.0	81.5	86.5	84.0	86.5	84.0
2	85.5	82.5	86.5	84.0	87.5	85.5
3	85.5	82.5	89.5	87.5	88.5	86.5
5	86.5	84.0	89.5	87.5	89.5	87.5
7.5	88.5	86.5	91.0	89.5	90.2	88.5
10	89.5	87.5	91.7	90.2	91.7	90.2
15	90.2	88.5	93.0	91.7	91.7	90.2
20	91.0	89.5	93.0	91.7	92.4	91.0
25	91.7	90.2	93.6	92.4	93.0	91.7
30	91.7	90.2	94.1	93.0	93.6	92.4
40	92.4	91.0	94.1	93.0	94.1	93.0
50	93.0	91.7	94.5	93.6	94.1	93.0
60	93.6	92.4	95.0	94.1	94.5	93.6
75	93.6	92.4	95.0	94.1	94.5	93.6
100	93.6	92.4	95.4	94.5	95.0	94.1
125	94.1	93.0	95.4	94.5	95.0	94.1
150	94.1	93.0	95.8	95.0	95.4	94.5
200	95.0	94.1	95.8	95.0	95.4	94.5
250	95.0	94.1	95.8	95.0	95.4	94.5
300	95.4	94.5	95.8	95.0	95.4	94.5
350	95.4	94.5	95.8	95.0	95.4	94.5
400	95.8	95.0	95.8	95.0	95.8	95.0
450	95.8	95.0	96.2	95.4	96.2	95.4
500	95.8	95.0	96.2	95.4	96.2	95.4

(continued)

TABLE 3.1   *(Continued)*

HP	ENCLOSED MOTORS					
	2 POLE		4 POLE		6 POLE	
	Nominal Efficiency	Minimum Efficiency	Nominal Efficiency	Minimum Efficiency	Nominal Efficiency	Minimum Efficiency
1	77.0	74.0	85.5	82.5	82.5	80.0
1.5	84.0	81.5	86.5	84.0	87.5	85.5
2	85.5	82.5	86.5	84.0	88.5	86.5
3	86.5	84.0	89.5	87.5	89.5	87.5
5	88.5	86.5	89.5	87.5	89.5	87.5
7.5	89.5	87.5	91.7	90.2	91.0	89.5
10	90.2	88.5	91.7	90.2	91.0	89.5
15	91.0	89.5	92.4	91.0	91.7	90.2
20	91.0	89.5	93.0	91.7	91.7	90.2
25	91.7	90.2	93.6	92.4	93.0	91.7
30	91.7	90.2	93.6	92.4	93.0	91.7
40	92.4	91.0	94.1	93.0	94.1	93.0
50	93.0	91.7	94.5	93.6	94.1	93.0
60	93.6	92.4	95.0	94.1	94.5	93.6
75	93.6	92.4	95.4	94.5	94.5	93.6
100	94.1	93.0	95.4	94.5	95.0	94.1
125	95.0	94.1	95.4	94.5	95.0	94.1
150	95.0	94.1	95.8	95.0	95.8	95.0
200	95.4	94.5	96.2	95.4	95.8	95.0
250	95.8	95.0	96.2	95.4	95.8	95.0
300	95.8	95.0	96.2	95.4	95.8	95.0
350	95.8	95.0	96.2	95.4	95.8	95.0
400	95.8	95.0	96.2	95.4	95.8	95.0
450	95.8	95.0	96.2	95.4	95.8	95.0
500	95.8	95.0	96.2	95.4	95.8	95.0

TABLE 3.2   Nominal efficiencies for "NEMA Premium®" induction motors rated medium volts 5 kV or less (form wound) (Tables reprinted by permission of the National Electrical Manufacturers Association, Rosslyn, VA. NEMA MG1-2006. NEMA Premium®)

	OPEN MOTORS					
	2 POLE		4 POLE		6 POLE	
HP	Nominal Efficiency	Minimum Efficiency	Nominal Efficiency	Minimum Efficiency	Nominal Efficiency	Minimum Efficiency
250	94.5	93.6	95.0	94.1	95.0	94.1
300	94.5	93.6	95.0	94.1	95.0	94.1
350	94.5	93.6	95.0	94.1	95.0	94.1
400	94.5	93.6	95.0	94.1	95.0	94.1
450	94.5	93.6	95.0	94.1	95.0	94.1
500	94.5	93.6	95.0	94.1	95.0	94.1

	ENCLOSED MOTORS					
	2 POLE		4 POLE		6 POLE	
HP	Nominal Efficiency	Minimum Efficiency	Nominal Efficiency	Minimum Efficiency	Nominal Efficiency	Minimum Efficiency
250	95.0	94.1	95.0	94.1	95.0	94.1
300	95.0	94.1	95.0	94.1	95.0	94.1
350	95.0	94.1	95.0	94.1	95.0	94.1
400	95.0	94.1	95.0	94.1	95.0	94.1
450	95.0	94.1	95.0	94.1	95.0	94.1
500	95.0	94.1	95.0	94.1	95.0	94.1

To be considered energy efficient, the motor's performance must equal or exceed the nominal full-load efficiency values provided by this standard. Specific full-load nominal efficiency values are provided for each horsepower, enclosure type, and speed combination. The motor's performance must equal or exceed the efficiency levels given in this standard.

The Energy Policy Act of 1992 requires that most general purpose motors manufactured for sale in the United States now meet this standard. This specifically covers 1200 hp (900 kW) general purpose, T-Frame, single-speed, foot-mounted, continuous-rated, polyphase, squirrel cage induction motors conforming to NEMA Design A and B. Covered motors are designed to operate with 230 or 460 V power supplies, have open or closed enclosures, and operate at speeds of 1200, 1800, or 3600 rpm (revolutions per minute).

7.7 Institute of Electrical and Electronic Engineers

7.7.1 Test Standards for Energy-Efficient Motors

The NEMA standard MG-1 recognizes the motor efficiency testing protocol established by IEEE 112 (IEEE, 2004) as the basis of testing for energy-efficient motors. The IEEE 112 standard test procedure recognizes five methods for determining motor efficiency. Each method bears certain advantages over the others with respect to accuracy, cost, and simplicity. The NEMA MG-1 references IEEE 112 method B. This test uses a dynamometer to measure motor output under load. Different testing methods yielding slightly different results are used in other countries. The NEMA nameplate labeling system for design of A and B motors in the 10 to 500 hp (7 to 370 kW) range uses bands of efficiency values based on IEEE 112 testing.

7.8 Consortium for Energy Efficiency

Consortium for Energy Efficiency is a nonprofit organization that includes many electrical utilities among its members. This group recognizes NEMA Premium motors up to 200 hp (150 kW) as meeting their criteria for possible energy-efficiency rebates.

The lack of labeling indicating a NEMA Premium motor does not necessarily mean the motor does not meet a high-efficiency standard. Some companies are not members of NEMA and therefore cannot apply the label. The owner may use other means to determine motor efficiency. Federal agencies are required to purchase NEMA Premium motors where applicable.

7.9 MotorMaster+

MotorMaster+ 4.0 is a software program developed through DOE that analyzes motor and motor system efficiency. MotorMaster+ 4.0 contains a database of over 12 000 motors available in the U.S. market. It also contains information on metric motors. Designed for utility auditors, industrial plant energy coordinators, and consulting engineers, MotorMaster+ 4.0 is used to identify inefficient or oversized facility motors and compute the energy and demand savings associated with selection of a replacement energy-efficient model. This free software is available through DOE (visit DOE's Web site at http://www1.eere.energy.gov/industry/bestpractices/software.html). However, the reader should note that the MotorMaster+ program has gone through several different updates. Please refer to the latest version in planning your energy program.

7.10 Motor Management

The process of energy efficiency does not stop with selection of the proper motor. The motor must be properly operated and maintained in order to continue to be energy efficient. Anticipating when a motor might fail can be important in achieving efficiency goals. Developing a sound motor management policy that determines when, whether, and how motors should be replaced and repaired is a necessary investment. Motor management also includes inventory tracking and maintenance of power quality. The DOE and corporations like Advanced Energy mentioned earlier in this chapter have motor management programs available.

7.11 Motor Records

Detailed specifications for all motors should be on file with maintenance records. Motor specifications found on the nameplate usually include the following: manufacturer's name, model number, serial number, rated power, voltage, number of phases, full-load amperage, speed, and service rating. The manufacturer of the motor should provide (separately) the motor efficiency and power factor at various percentages of the full-load rating together with recommended preventive maintenance requirements and replacement part catalog numbers.

The following are recommended best management practices:

- If requirement data are not on file, they should be obtained from the manufacturer.

- Before a motor is placed into service, the nameplate information must be recorded or checked against the design specifications. Motor nameplates become illegible with time because of paint, scraping to remove paint, oil, weather, submersion in wet wells, and other factors. Therefore, the only definite way to have the information when the motor must be serviced or replaced is to write it down when it is installed.

7.12 Motor Failure

Theoretically, a squirrel cage electric motor (when operated fully within its design parameters) should never fail because the only parts that wear are the wearing surfaces. In addition, a properly loaded modern bearing will run practically indefinitely, provided it receives proper care. The magnetic lines of flux that cause the motor to operate cause no wear; the electricity flowing through the wire does not wear out the wire. Therefore, the motor should never wear out and fail. This theory applies only to squirrel cage motors because the brushes and slip rings in the other types are subjected to wear, which will eventually cause motor failure.

The primary causes of motor failure are neglect, including failure to keep the motor vents clean, which results in overheating; improper lubrication (either too much or too little), which causes bearing failure and may result in complete destruction of the motor if the rotor comes in contact with the stator; and improper belt tension, which results in bearing failure and possible destruction of the motor. Failure to maintain proper overload relays and other protective equipment may cause an electrical overload. Heat and chemical action from excess oil cause breakdown of the insulation and shorted windings. Failure to maintain the driven machinery results in overloading, overheating, and, finally, motor failure. Lack of proper protection against environmental factors results in motor failure due to damage by rust, corrosion, or contamination from foreign particles. Surprisingly, many motors are damaged by water during the washdown process by careless WWTP personnel. Motor control circuits are often neglected until a failure occurs. Loose connections in wiring circuits cause overcurrent or phase failure as well as overheating. These are a few, but not all, of the forms of neglect that result in motor failure (WPCF, 1984). The design life of good bearings should be at least 100 000 hours (or 11.4 years of continuous duty). The primary cause of bearing failure is excessive lubrication.

The following are recommended best management practices for motors:

- Periodic lubrication by trained technicians is preferable to lifetime lubrication.

- All motor bearing casings should be fitted with a grease relief to permit exit of excess grease. If the grease relief is not open, the excess grease will cause high bearing temperatures and premature failure. Excess grease may also be forced inside the motor and coat the windings, causing them to overheat.

- The following items should be checked periodically to ensure long bearing life: external or internal grease accumulations, dirt buildup, excessive motor temperature (especially bearing ends), shaft alignment, bearing play (run-out), and excessive noise or vibration. Frequently checking the mechanical operation of a motor and driven equipment for vibration, noise, high temperature, or other abnormal conditions and providing proper routine maintenance of dust removal and bearing lubrication are extremely important for ensuring long motor life.

- The electrical operation of the motor affects the cost of operation and the ability of the motor to provide dependable service. The motor should be checked for proper voltage, amperage, and resistance to ground. How these measurements are made will be covered in detail in the following section. Operation of the motor controls should also be checked carefully. Motor starters should engage fully and not chatter. Weak coils and worn contacts should be replaced. Thermal sensors and cameras can be used to check for the proper dissipation of heat on the contacts and to verify the connection tightness. Once per year, all connections in the control panel should be checked for tightness.

7.12.1 Motor Repairs

The failure of any motor brings about several decisions. First there is the need to determine the exact cause of the failure. The continuation of a solvable problem defeats any energy savings solution.

Therefore, before ordering a direct replacement the owner needs to review the guidelines set for energy-efficient designs. It may be found that the new motor can be upgraded using the guidelines set for energy efficiency.

7.12.2 Selection of a Motor Repair Facility

All motor repair facilities are not equal. Bringing a damaged motor back to its "as-built" condition requires a level of skill that is not available in all repair shops. A poorly repaired motor can easily be a less efficient motor, thus defeating the plant's energy efficiency program. Before committing to a specific repair, an audit of the motor repair facility should be conducted as follows:

- Determine the condition of the equipment that will be used for the repairs.
- Determine if proper safety standards for shop operation are in effect.
- Determine if proper record keeping is maintained.
- Determine the knowledge level of the staff involved with the repair.
- Determine that the repair facility stocks and uses the proper wire and components that are needed.
- Check the test equipment.
- Check if the repair facility complies with motor repair quality standards or equipment specifications.

- Check if the facility complies with standards and procedures published by the Electrical Apparatus Service Association.
- Determine if the facility has a written quality program.

7.13 Transformers

A transformer is a device that transfers electrical energy from one circuit to another through inductively coupled electrical windings. Transformers in water or wastewater plants are primarily used to increase or decrease the voltage level or for isolation. This section will focus primarily on large transformers that have a greater effect on overall energy usage. Transformers have many similarities to electric motors in that they both are based on winding electrical wire around a core. An example of a transformer circuit is illustrated in Figure 3.9.

As with electric motors, the efficiency of the transformer is directly related to the density of winding, amount of copper, spacing, and selection of materials. Although standard transformers have historically been efficient devices, typically between

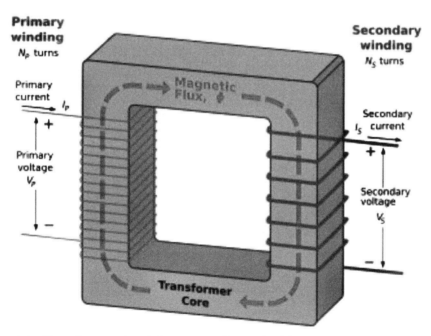

FIGURE 3.9 Transformer circuit (courtesy of LearnHub).

95 to 96.7%, there has been room for improvement. This need for energy savings has led to new standards. Transformer losses are attributable to several causes and may be differentiated between those originating in the windings, sometimes termed *copper loss,* and those arising from the magnetic circuit, sometimes termed *iron loss.* The losses vary with load current, and may furthermore be expressed as "no-load" or "full-load" loss, respectively. Winding resistance dominates load losses, whereas hysteresis and eddy current losses contribute to over 99% of the no-load loss. The no-load loss can be significant, meaning that even an idle transformer constitutes a drain on an electrical supply and lending impetus to development of low-loss transformers.

The NEMA TP-1 standard addresses medium- and low-voltage transformers (NEMA, 2002). This standard defines energy efficiency levels of 97.9 to 99.0%. Preliminary analyses indicate a 20 to 40% reduction in transformer losses with TP-1 equipment. The use of an energy-efficient NEMA TP-1 transformer has the potential of reducing annual electrical costs from 0.5 to 3.0%. As with NEMA Premium electric motors, the engineer and owner need to evaluate these products based on energy savings over their expected life. In almost all cases, this strongly favors the use of energy-efficient transformers.

8.0   REFERENCES

Institute of Electrical and Electronics Engineers (2004) *Standard Test Procedure for Polyphase Induction Motors and Generators;* IEEE 112-2004; IEEE Press: Los Alamitos, California.

National Electrical Manufacturers Association (2006) *American National Standard for Electrical Power Systems and Equipment—Voltage Ratings (60 Hertz);* ANSI/IEEE C84.1; National Electrical Manufacturers Association: Rosslyn, Virginia.

National Electrical Manufacturers Association and American National Standards Institute (2006) *Motors and Generators;* ANSI/NEMA MG-1-2006; National Electrical Manufacturers Association: Rosslyn, Virginia.

National Electrical Manufacturers Association (2002) *Guide for Determining Energy Efficiency for Distribution Transformers;* NEMA TP-1; National Electrical Manufacturers Association: Rosslyn, Virginia.

Water Pollution Control Federation (1984) *Prime Movers: Engines, Motors, Turbines, Pumps, Blowers & Generators;* Manual of Practice No. OM-5; Water Pollution Control Federation: Washington, D.C.

9.0 SUGGESTED READINGS

ITT FLYGT (1985) FLYGT Product Education Manual, 8th ed.; Product Education: Norwalk, Connecticut.

Polka, D. (2003) *Motors & Drives: A Practical Technology Guide;* International Society of Automation: Research Triangle Park, North Carolina.

U.S. Environmental Protection Agency (2008) *Ensuring a Sustainable Future: An Energy Management Guidebook for Wastewater and Water Utilities.* http://www.epa.gov/waterinfrastructure/pdfs/guidebook_si_energymanagement.pdf (accessed Feb 2009).

Water Environmental Federation (2007) *Automation of Wastewater Treatment Facilities,* 3rd ed.; Manual of Practice No. 21; Water Environment Federation: Alexandria, Virginia.

Washington State University Energy Program (2003) Energy Efficiency Facts. http://www.energy.wsu.edu/documents/building/light/compact_fluor.pdf (accessed Feb 2009).

Chapter 4

Pumping

Pumps represent significant energy uses in wastewater collection and water distribution systems. Whereas most features of pump system efficiency are determined during design and construction, there are numerous operation and maintenance (O&M) practices that can be implemented to either improve pumping system efficiency or restore original system efficiency. However, it is impossible to simply look at a pump and determine whether it is performing in an energy-efficient manner. Inefficient pumping operations typically do not smell bad, sound alarms, give off smoke, or fill customers' basements with wastewater—they simply waste energy and money. Calculations, field measurements, analyses, and research data recorded in pump curves are required to determine whether a pump is operating efficiently and evaluate what can be done to conserve energy. Because considerable effort is required to locate the source of pump inefficiency, improving pumping efficiency is often a neglected issue. This chapter focuses on the type of pumping one would find at pump stations as opposed to sludge pumping, where one is more likely to use positive displacement pumps.

1.0 PUMPING PRINCIPLES

Pumps are used in wastewater collection systems to raise water to high points in the system, to which the water cannot flow by gravity, or pump wastewater long distances through flat terrain. Even where gravity flow is possible, it is sometimes more economical to pump water through a force main because a force main can be considerably smaller than a gravity main of the same capacity. In addition, pumps are used to raise wastewater from sewers that flow into most wastewater treatment plants (WWTPs), enabling it to flow by gravity through the plant.

In water distribution systems, pumps are used to pressurize the system and raise water to higher pressure zones. Pumps are also used to deliver raw water from a surface or groundwater source to the water treatment plant. Numerous references are available on water and wastewater pumps and on energy conservation (see the "Suggested Readings" section at the end of this chapter.)

Energy consumption for a pump is dependent not only on the pump, but on the hydraulics of the system in which it is installed. A pump may be rated at 2700 m³/d (500 gpm) at 18 m (60 ft) of head with 70% efficiency, but it will not discharge 2700 m³/d (500 gpm) at 18 m (60 ft) and 70% efficiency unless the hydraulics of the suction and discharge sides of the pump are matched exactly with the pump. Any deviation from those conditions will mean that the pump will perform less efficiently even though it may actually produce more discharge or more head.

To determine the actual discharge, head, and efficiency of a pump, it is necessary to understand two types of curves: pump characteristic curves and system head curves. These curves can be used to determine the operating point of the pump—the discharge, head, and efficiency.

1.1 Pump Characteristic Curves

Pump characteristic curves received from the manufacturer are a reflection of the pump in new condition and are independent of the installation. Pump characteristic curves are plots of pump properties such as head, efficiency, power required, and net positive suction head required versus discharge. A typical set of head and efficiency curves is shown in Figure 4.1a and is relevant to a single pump with a single-sized impeller operating at a single speed. Because a given pump can usually accommodate a range of different impeller sizes, pump characteristic curves prepared by a manufacturer often show a family of curves for several impeller sizes, as shown in Figure 4.2.

The most important pump characteristic curves for energy conservation are the head characteristic and efficiency characteristic curves.

There are two important points on these curves: the best efficiency point (BEP) and the operating point. The BEP is described in the following section. To understand the operating point, it is necessary to understand system head curves; as such, a description of the operating point follows the discussion of system head curves.

1.2 Best Efficiency Point

The BEP on a pump curve is the point corresponding to the maximum efficiency of the pump. For example, in Figure 4.1a, the BEP corresponds to a flow of 5500 m^3/d (1000 gpm) with a head of 30.5 m (100 ft). The rated capacity (or duty point) of the pump is the discharge and head at (or near) the BEP. The BEP is a property of the pump and motor alone.

1.3 System Head Curve

The actual performance of a pump in any situation depends on the hydraulic conditions it encounters at any point in time. For water and wastewater pumps, this depends on a number of factors, including:

- Water level (or pressure) on the suction side of the pump relative to the pump impeller centerline;

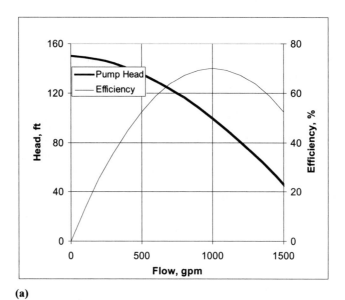

(a)

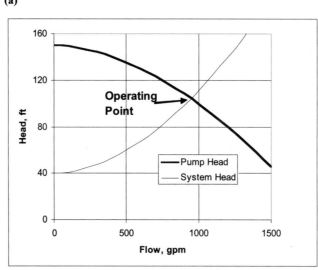

(b)

FIGURE 4.1   Typical pump characteristic curves.

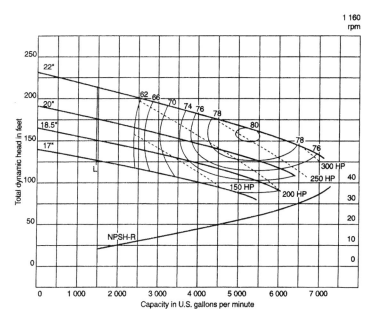

FIGURE 4.2 Pump characteristic curves for several impeller sizes (ft × 0.304 8 = m; gpm × [6.308 × 10^{-5}]) = m^3/s.

- Water level on the discharge side of the pump, if the pump discharges at or below the water surface (or the elevation at which the flow changes from full pipe to partly full on a free discharge). For water systems, this corresponds to the water level in the elevated tanks receiving the pumped water and, for wastewater pumps, this corresponds to the water level in the manhole or tanks where flow converts back from full pipe to partly full;

- Head loss in the suction and discharge piping, which depends on the flows, diameters, and roughness of the pipes; and

- Influence of other pumps discharging into the same discharge line or pressure zone (whether they are in the same pumping station or other pumping stations) and location and demand of water users on the discharge side of the pump for water systems.

The relationship between the discharge of a pump and the head it must pump against is called the system head curve. In practice, there is not a single system head curve but, rather, a band of system head curves because the water levels in tanks, valve settings, customer demands, and operation of other pumps can and do change over time. For example, tank levels rise and fall, other pumps turn on and off (or change speeds, in the case of variable-speed pumps), valves can be throttled or closed, and pipe roughness and deposits can increase over time.

When tank levels remain constant and the status of other pumps in the same discharge line do not change, the system head curve can be represented by a single curve. The curve is always upwardly sloping because, as pump discharge increases, head loss resulting from friction in the pipes also increases. For simple situations, the system head curve is the sum of the head required to lift the water from the suction tank to the discharge tank plus the head needed to overcome friction losses, as follows:

$$h = h_l + h_f \tag{4.1}$$

Where

h = system head, m (ft);

h_l = lift between suction and discharge point, m (ft); and

h_f = friction loss between suction and discharge point, m (ft).

The lift is easy to determine by operational measurements or from engineering drawings of the system. For simple pipelines, the friction loss can be determined using an equation such as the Hazen-Williams equation, Darcy-Weisbach equation, or the Manning equation. For complicated pipelines, especially those with several pumping stations on the same discharge line, a computer hydraulic model may be required to predict friction loss for a variety of conditions. In the field, points on the system head curve can be determined using pressure gauges or transducers attached to the suction and discharge side of the pump or a differential pressure gauge or transducer connected to each side of the pump. More information on system head curves is available from Ormsbee and Walski (1989).

Some typical system head curves are shown in Figure 4.3. In general, the lift term in Eq. 4.1 will dominate for water distribution pumping and wastewater force mains pumping over high points, whereas the friction loss term dominates for long force mains in relatively flat terrain. The nonlinear nature of the head loss equations results in a relatively flat system head curve for low flowrates with an increasing slope as flow increases to the capacity of the piping.

In determining the strategy to be used for energy conservation, it is helpful to distinguish between different extreme types of system head curves. On one extreme is a flat system head curve (curve A in Figure 4.3), which is typical for systems with short, large pipelines and little friction loss, such as those found at influent pumping stations in WWTPs or in systems with pipelines that have been sized with considerable excess capacity. On the other extreme are steep system

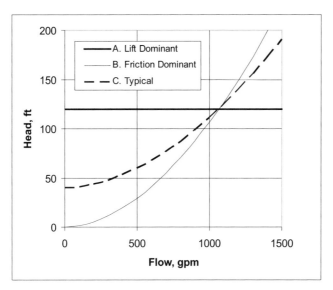

FIGURE 4.3   Examples of system head curves.

head curves (curve B in Figure 4.3), which are characteristic of systems with long force mains, high velocity, high friction losses, and little or no lift, typical of booster pump stations in flat areas. Most system head curves lie somewhere between the two extremes. Energy conservation measures are considerably different for each extreme, as follows:

- For curve A, raising the wet well level can save some energy (for curve B, there would be little improvement); and

- For curve B, cleaning the discharge line by "pigging" or scraping can be helpful (cleaning would result in no improvement for curve A).

For multiple pumps discharging into a common force main manifold, system head curves can vary widely. Figure 4.4 shows how system head curves can vary depending on whether other pumps in the same force main are running and, if variable-speed, whether they are running at high or low speed. During wet weather or other peak times, when it is likely that most other pumps are running simultaneously, a pump will need to discharge against a higher head than during low-flow-off-peak times. Consequently, pump efficiency and energy use will vary depending on the system head developed at the given time by the combination of pumps and resulting head, along with the pumps' characteristics.

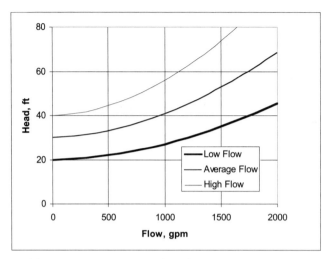

FIGURE 4.4 Possible variation in system head curves.

1.4 Operating Point

The operating point is the point corresponding to the actual discharge and head produced by the pump. It is the only point that satisfies both the pump properties and the system properties. Graphically, it is the intersection of the pump head characteristic curve and the applicable system head curve at that point in time. The operating point depends on the pump and the resulting hydraulics of the system in which it is installed; it is shown graphically in Figure 4.1b.

1.5 Relationship of Best Efficiency Point and Operating Point

Ideally, the operating point will be close to the BEP. Because the system head curve can continuously change, this will not always be possible. Therefore, pumps should be selected and operated so that the operating point is reasonably close to the BEP.

When the BEP is not located near the operating point, it is necessary to investigate the cause of the difference. If the pump is not operating on its characteristic curve, air in the piping, incorrect clearances, worn bearings, or even improper impeller size or motor speed are possible explanations. Typically, however, pumps perform on or near their characteristic curves.

If a pump is operating on its characteristic curve to the left (low-flow side) of its BEP, it is producing less flow than is optimal. The pump is experiencing much more resistance to flow than it can handle. This can be the result of an undersized discharge line, some type of blockage (e.g., partly closed valve) or encrustation in the suction or discharge piping, higher levels in the discharge tank (or lower levels in the suction tank) than anticipated, or a higher head in the discharge line than anticipated because of the operation of other pumps discharging to the same discharge line. The remedy is to reduce the resistance the pump must overcome or increase the head the pump produces.

If the pump is operating on its characteristic curve to the right (high-flow side) of the BEP, it is producing more flow than anticipated because it is experiencing less resistance to flow than anticipated. This can be because of an oversized discharge line, discharge or suction lines being cleaner than anticipated, low water level in the discharge tank (or high water level in the suction tank), or lower head in the discharge line than anticipated because of the operation of other pumps. Changing the impeller or pump operation may remedy this problem.

Regardless of the reason, the extent to which a pump operates away from its BEP is a reflection of the extent to which the pump is mismatched with the hydraulics of the system it serves. This can be the result of an improper selection of pumps, a change in operating conditions after the pump was installed, or simply an expectation that the pump will operate over an excessively wide range of conditions.

2.0 ENERGY PRINCIPLES

Energy refers to the ability to do work and produce power. Pumps convert some form of energy (mechanical, electrical) into water energy. The power required is defined in terms of work as weight of water per unit time (force) multiplied by total dynamic head (distance). A commonly used formula defining water power is

$$p_w = K Q h \qquad (4.2)$$

Where

p_w = water power, kW (hp);

Q = flowrate, m^3/s (cfs, gpm, mgd);

h = pump head, m (ft); and

K = unit conversion factor

9.8 for kW, m^3/s, m

1.1×10^{-4} for kW, m^3/d, m

0.113 for hp, cfs, ft

2.53×10^{-4} for hp, gpm, ft

0.175 for hp, mgd, ft.

2.1 Pump Efficiency

The key to successful energy conservation for pumps is to produce water power using the smallest amount of electrical (or other) power as possible. The ratio of water power produced by the pump to its power input is known as *overall pump efficiency* or *wire-to-water efficiency*. The overall efficiency must consider the inefficiencies of the pump, the drive or transmission, and the power source such as motor or engine.

The *mechanical efficiency* of the pump is the ratio of water power from the pump to mechanical power input to the pump shaft. In formulaic terms for a direct-drive coupled motor:

$$e_p = \frac{K\,Q\,h}{p_m} = \frac{\text{water power}}{\text{motor power}} \tag{4.3}$$

Where

    e_p = pump efficiency; fraction;
    Q = flowrate;
    h = pump head; and
    p_m = output of the motor, kW (hp).

It is important to note that motors are rated by output rather than input. Therefore, a 75-kW (100-hp) motor should deliver the full 75 kW (100 hp) to the transmission or, in direct-drive applications, to the pump.

The motor and any drive mechanisms are not perfectly efficient. The efficiency of the motor can be determined as follows:

$$e_m = \frac{p_m}{p_e} = \frac{\text{motor power}}{\text{input power}} \tag{4.4}$$

Where

    e_m = motor efficiency and
    p_e = electric power (motor input power), kW (hp).

The efficiency of any associated drives (e.g., variable-frequency drives [VFDs]) must also be considered using a similar formula as follows:

$$e_d = p_p/p_m \tag{4.5}$$

Where

    e_d = transmission efficiency and
    p_p = pump input power (or transmission output power), kW (hp).

The overall efficiency of the pump and motor, referred to previously as the wire-to-water or overall efficiency, is the product of the pump efficiency, drive efficiency, and motor efficiency, as follows:

$$e_{ww} = e_m \times e_d \times e_p \qquad (4.6)$$

Where

e_{ww} = wire-to-water efficiency.

While efficiency is actually a number between zero and one, it is typically expressed as a percentage (that is, 80% rather than 0.80). In equations presented later in this chapter, the decimal form of the efficiency is used.

Typically, when pumps are tested by the manufacturer, the mechanical power input to the pump is controlled, and the resulting curves of efficiency versus flow are pump efficiency curves. In field applications, it is difficult to measure motor power input to the pump, so field testing of pumps almost always involves measurement of wire-to-water efficiency.

For example, a pump uses 4.0 MJ (1.1 kWh) of electricity in 10 minutes to pump 2200 m³/d (400 gpm) against a head of 17 m (55 ft). To determine the wire-to-water efficiency of the pump and motor, calculate water power as follows:

$$Q(\text{cfs}) = \frac{400 \text{ gpm}}{(448 \text{ gpm/cfs})} = 0.89 \text{ cfs} = 0.025 \text{ m}^3/\text{s}$$

hp (water) = [(62.4 lb/cu ft)(0.89l cfs)(55 ft)]/(550 ft-lb/sec-hp) = 5.56 hp = 4.1 kW

Calculate electric horsepower as follows:

hp (electric) = (1.1 kWh × 60 min/hr)/(0.746 kW/hp)/(10 min) = 8.85 hp = 6.59 kW

$$e_{ww} = 5.56/8.85 = 0.63, \text{ or } 63\%$$

2.2 Energy Consumption

The energy consumption over a period of time is the water power multiplied by the length of time and divided by the wire-to-water efficiency:

$$E = \text{hp (water)} \times 0.746 \text{ kW/hp} \times t/e_{ww} \qquad (4.7)$$

Where

E = energy, kWh, and

t = time over which energy is consumed, hours.

The cost of energy can be determined by multiplying the energy used or forecast by the price of energy, as follows:

$$C = \frac{E \times P}{100} \qquad (4.8)$$

Where

C = cost of energy during time t, and

P = price of energy, cents/kWh.

Substituting Eq. 4.7 for E,

$$C = \frac{[\text{p (water)} \ kW \times t / e_{ww}] \times P}{100} \qquad (4.9)$$

Substituting Eq. 4.2 for horsepower (water),

$$C = KQH \ Pt / e_{ww} \qquad (4.10)$$

If K corresponds to power in horsepower, then Eq. 4.10 needs to be multiplied by 0.745. If the price of energy changes over time (e.g., time-of-day energy pricing), then the cost of energy will need to be calculated separately for each time period and summed.

To determine the annual cost of energy for a given pump, the number of hours that pump ran during the year is substituted for t. The energy use and cost of energy can be summed over all the pumps in a station to determine the energy cost for a year (or any other time period).

Find the expected annual energy cost to operate a raw water intake pump at a water treatment facility that, on average, pumps 1.1 m³/s (37 cfs) (24 mgd) with 12 m (40 ft) of head, has an overall average efficiency of 60%, runs all year (8760 hours), and has an energy cost of 8 cents/kWh:

Annual Cost = 9.8(1.1 m³/s)(12 m)(8 cents/kWh)(8760 hr)/(0.60 × 100) = $151,000/yr.

2.3 Multiple Operating Points

In most cases, it is not sufficiently accurate to use the average flow, head, and efficiency of a pump (or station) to calculate energy cost. Instead, the values corresponding to each operating point should be calculated and summed. If there is an operating point possible for a pump or group of pumps at a pumping station during time period of length T, then energy cost can be calculated as:

$$C = 0.01\,P\sum_{i=1}^{n} p_i\, t_i\, /\, e_{wwi} \qquad (4.11)$$

Where

$$\sum_{i=1}^{n} t_i = T \qquad (4.12)$$

i = subscript of operating point,

n = number of operating points,

p_i = power used at i-th operating point, kW,

t_i = time spent at i-th operating point, hr, and

e_{wwi} = efficiency at i-th operating point.

For example, a pumping station can operate with zero, one, two, or three identical pumps at any time. During the course of a year, the station operates as shown in the first five columns of Table 4.1. Find the energy cost if the price of power is 9.2 cents/kWh.

As a quick approximation, round off and assume that, on average, the pumping station pumps 0.069 m³/s (1100 gpm), with 14 m (46 ft) of head and an efficiency of 61%, and use Eqs. 4.2 and 4.11 to give:

Annual Cost = 2.53×10^{-4} (1100)(46)(9.2)(8760)(0.745)/61 = $12,586 ~ $12,600

TABLE 4.1   Energy cost example for pumping stations

Pumps running	Q_i, gpm	h_i, ft	t_i, hr	e_i, %	p_i, kW	Cost$_i$, $
0	0	0	240	—		
1	900	45	6260	62	12	7091
2	1700	48	2120	61	25	4918
3	2500	55	140	57	45	586
Sum						12 594

1 gpm = 5.54 m³/d, 1 ft = 0.305 m

2.4 Measuring Pump System Efficiency

Utilities seldom check the efficiency of in-place pumping stations even though such checks can identify large inefficiencies and point to areas where energy can be saved. Because of the difficulty measuring motor power input to a pump, it is usually much easier to check the overall (wire-to-water) efficiency of a pump. Remembering that efficiency is the ratio of power out to power in, the efficiency can be calculated by measuring flow, head, and power input. This can be accomplished with a flow meter, two pressure gauges, and a method to measure power input as follows:

$$e_{ww} = \frac{KQh}{p_e} \tag{4.13}$$

Where
$\quad Q$ = flow,
$\quad h$ = head, and
$\quad p_e$ = power input.

The power input can be measured either directly with a power meter, with an ammeter converted to power units knowing the voltage and phase angle, or by clocking the cumulative energy meter at the station over the duration of the measurement and dividing by time. Some VFDs can display power input and output.

For example, consider a pump station producing 4370 m³/d (800 gpm), with suction and discharge pressures of 69 kPa (10 psi) and 427 kPa (62 psi) (h = 2.31 [62 − 10] = 120 ft = 36.5 m) while using 27 kW. The overall efficiency is:

$$e_{ww} = \frac{KQh}{p} = \frac{1.1 \times 10^{-4}(4370)(36.5)}{27} = 0.65 = 65\%$$

2.5 Demand Charge

The cost of energy is not the only charge that is levied on the pumping station. There is usually another charge based on the peak power consumption during some time period. This charge is usually called the *demand charge* or *capacity charge* and reflects the cost to the power company to provide sufficient capacity to meet the pumps' peak demand. Tariffs vary widely between power companies. Some are based on the maximum 15 minutes of power use during the previous month or may be based on the maximum 1 minute use during the previous year. In some cases, it is based on the pump stations' power use during a power company's peak demand period.

Because of this charge, it is important for operators to avoid pumping at a high rate; rather, they should even out the pumping over time. Unfortunately, operators do not have complete control over peak pumping rates. In wastewater utilities, the demand charge is usually established during a day where a great deal of wet weather flow occurs. For water utilities, it is usually set on a hot, dry day or a day when a large fire occurs. It is important to make operators aware of the impact of demand charges so that they do not accidentally set the peak power demand higher than it needs to be. For more information on energy pricing, refer to Chapter 2, "Utility Billing Procedures and Incentives."

3.0   REDUCING ENERGY USE AND COST

To reduce energy cost, it is useful to look at the energy equation in Eq. 4.7 described earlier in this chapter and see the terms that influence energy use and costs. The overall energy cost equation can be written as follows:

$$\text{Energy Cost} = K \frac{Qhp}{e_p e_d e_m} \tag{4.14}$$

Where K is a constant depending on the units (see Eq. 4.2 for a definition of the other quantities). Each of these quantities (flow, head, price, and the three efficiencies) can be manipulated to increase or decrease energy costs. The methods to improve each of these quantities will be described and discussed in the following sections.

   In general, before implementing any energy use/cost reduction methods, it is best to simulate the results using an extended period simulation hydraulic model coupled with energy cost calculations. This will help estimate the expected cost savings and identify any areas that might be adversely affected by the method.

   When comparing different ways of designing and operating systems, it is important to not simply compare efficiency or costs over a short period of time. Instead, it is important to compare energy costs over a long enough period of time (e.g., at least a day). Instead of comparing based on total costs, it is often useful to compare energy costs on a cost-per-volume-pumped basis (e.g., dollars per million gallons pumped). Where water use and wastewater generation significantly varies between seasons, it is also useful to compare energy costs during different seasons.

4.0   DISCHARGE

Reducing pump discharge is the most direct way to reduce energy consumption. Energy savings as a result of reductions in flow are typically proportional to the amount of flow reduced, and sometimes even greater. There are often opportunities to realize savings by minimizing the flow in wastewater systems, for example, by reducing recycle streams and infiltration and inflow into the collection system. In water systems, this can be accomplished by reducing leakage.

For constant-speed pumps discharging to a manhole or elevated storage tank, reductions in flow typically are realized at a pumping station as a reduction in the amount of time the pumps actually run, rather than appearing as a reduction in instantaneous flowrate while the pumps are running. This means that for a single pump on a pipeline, energy savings should be proportional to flow reduction.

For variable-speed pumps, reduction in flow to the pumping station appears as a decrease in the flow and head produced by the station. It will rarely reduce the duration for which the pumps run. Depending on the pump efficiency curve, reductions in flow may result in decreased average pump efficiency. Nevertheless, the major savings will be from the reduced flow.

In water distribution systems, the volume pumped can be reduced by detecting and correcting leaks in the system. The savings should be roughly proportional to the reduction in leakage. Depending on the local situation, energy cost savings alone can be a major portion of the justification for leak detection programs.

Another source of energy savings is not pumping water into a higher pressure zone only to feed it back into a lower zone through a pressure-reducing valve. Although systems are typically not planned to operate this way, these situations can evolve over time, especially in areas with hilly terrain. Laying out the system so that pumps will not need to discharge to a high-pressure zone only to have the water flow back down to a lower zone results in savings directly proportional to the water that would have passed from the upper to the lower zone.

5.0 HEAD

Reducing the head that a pump must work against results in a proportional reduction in energy consumption, provided the efficiency of the pump and motor do not change significantly. Because head is the result of lifting water or overcoming friction, reduction in head can be achieved by reducing the lift or minimizing friction losses.

5.1 Measuring Pump Head

The best way to measure the head that a pump is producing is to measure the head with a differential pressure gauge with one tap immediately upstream and the second tap immediately downstream of the pump. For greatest precision, the gauge should have a range slightly greater than the anticipated shutoff head of the pump. The change in velocity head should be added or subtracted from the pump head reading. In most cases, however, the change in velocity head across the pump is negligible compared to differences in pressure head.

If a differential gauge is not available, then separate pressure gauges for the discharge and suction side of the pump should be used. In some instances, the suction gauge may need to be a compound pressure-vacuum gauge. If the discharge gauge is at a higher/lower elevation than the suction gauge, the difference in elevation of the gauges should be added to/subtracted from the pump head. If the pump is a submersible pump, then there is no need to measure suction pressure (which is atmospheric), and the pump head is simply the discharge pressure converted to head units plus the difference in elevation between the suction water level and the pressure gauge.

Occasionally, a convenient location to tap the pipe near the pump is not available and it becomes necessary to use a pressure reading at a more distant location to determine the suction or discharge pressure. In such a situation, it is necessary to estimate the head loss in pipe and fittings between the measured head and the pump and the change in velocity head and subtract these heads from the measured suction head (or add these values to the measured discharge head). The minor losses in the station are usually small compared to the pump head.

5.2   System Head

In instances where a single pumping station discharges to a force main, there is little that can be done to reduce the discharge head. However, controlling the flow with variable-speed pumps and minimizing multiple-pump or peak flow operation will minimize head losses resulting from friction, which is especially important in systems with long mains.

When multiple pumping stations discharge into a force main, it is important to ensure that the pumps are evenly matched. Otherwise, one or two pumps with high heads can increase pressures in the force main so that the higher head pumps back up and possibly even hydraulically "shut off" the lower head pumps (shutoff head is the point at which the pump characteristic curve intersects the vertical axis, i.e., discharge equals zero). Because all of the wet wells will be at slightly different elevations, it is important to make these comparisons based on discharge hydraulic grade line relative to the same datum point (e.g., sea level) and not simply make comparisons based on discharge head or pressure. This is especially true for constant-speed pumps, but problems can even be experienced in variable-speed stations.

Analysis of force mains with multiple pumps is best done by running a computer hydraulic model of the force main system and looking at the pump operating points during high, low, and average conditions. For example, the results may show that during low flow, some variable-speed pumps are running so slowly and inefficiently that it is better to simply turn them off until the wet well level rises. Whereas in other instances during high flow, some pumps may not produce any flow at all.

For example, given that there are four pumping stations discharging to a force main, determine the discharge head for each and determine whether they will work well together. Data on the pumping stations are given in the first four columns of Table 4.2. The last two columns are the sum of the average wet well level and the appropriate pump head. Within each pumping station, all pumps are identical.

In this example, pumping station 3 is not well matched with the other stations. It would tend to use up the capacity of the force main and, if the force main does not have a large capacity, it may cause the other pumping stations to run inefficiently. If the force main has adequate capacity and pumping station 3 has variable-speed drives, the combination in the example may be acceptable. However, it is best to avoid such a situation.

In water distribution systems, the pump head is controlled by the difference in water level in the suction and discharge side tanks or pressure in the suction and discharge piping. An appropriate water level needs to be maintained in tanks for pressure equalization, emergencies, and fire fighting, but the tank water level should be allowed to fluctuate for water quality turnover. A side benefit of the turnover in the tank is that for some of the time, the pump is discharging into a system with a lower head.

In a closed water distribution system with no discharge side storage and variable-speed pumping, the setpoint for the variable-speed pump can be reduced to

TABLE 4.2   Data for four pumping stations discharging to a common force main

Pump station	Average wet well elevation m[b]	Pump head at BEP[a] m	Shutoff head m	Force main head at BEP m	Force main head at shutoff m
1	159	14	18	172	177
2	154	17	23	171	177
3	152	24	34	177	186
4	149	20	26	169	17

[a] Best efficiency point
[b] ft × 0.3048 = m

reduce the pump head. Before doing so, it is advisable to study how the pressures in the system will respond using a hydraulic model to ensure that no one will receive inadequate pressure. In some instances, it may be acceptable to reduce the setpoint only during nighttime hours, which has the added benefit of reducing leakage during this period. In wastewater pumping, the pumps will work against less head if the levels maintained in the wet well are increased. This improvement is marginal and must be weighed against a longer detention time and reduced safety factor in case of a short-term power outage.

5.3 Friction

Friction losses are inevitable whenever fluid flows, but they can become excessive and result in increased energy use. Situations in which normal friction losses can increase include:

- Lack of capacity of the force main,

- Excessive roughness in the force main, and

- Blockage of the force main with debris or a partly closed valve.

5.4 Determining Head Loss

The friction energy loss is the sum of the loss in straight runs of pipe, which can be determined from the Hazen-Williams, Manning, or Darcy-Weisbach equations, plus minor losses resulting from valves, bends, and fittings, which can be determined from hydraulic handbooks or manufacturers' literature. For long pipelines, minor losses are usually negligible compared with line losses, but, for short lift lines at WWTPs, minor losses can be of greater significance.

The Hazen-Williams equation can be written as:

$$h = (K' \, L/D^{4.87})(Q/C)^{1.85} \qquad\qquad (4.15)$$

Where
h = head loss, m (ft);
L = length of pipe, m (ft);
D = diameter, m (ft, in.);
Q = flow, L/min (gpm);
C = Hazen-Williams C-factor; and
K' = 10.7 for D, L in m, Q in m^3/s
       4.71 for D, L in ft, Q in cfs
       10.4 for D in in., L in ft , Q in gpm.

The C-factor is an indicator of the carrying capacity (smoothness) of the pipe and can range from 140 for new, smooth pipe to 30 for old encrusted/tuberculated pipe. Values of around 100 are typically used for design, although this can result in a pump that provides more head than needed and possibly operates at an inefficient point.

Minor losses for valves and fittings can be converted to length of straight pipe, which can be added to actual lengths of straight pipe to estimate overall head loss. For example, for a 0.15-m (6-in) pipe with a fitting that would result in a loss of 6 diam, the fitting would cause the same head loss as 0.9 m (3 ft) of pipe. One would need to add 0.9 m (3 ft) to the length used in the straight pipe head loss formula to account for the losses caused by the fitting. Equivalent pipe-length conversion values for common valves and fittings can be found in many books, handbooks, and other references (e.g., American Society of Civil Engineers [1992]; Jones et al. [2008]; Walski [1984]; and Walski et al. [2003]).

A quick way to determine whether capacity is adequate is to check the velocity in the pipe during normal operation. The velocity in wastewater piping should be greater than 0.6 m/s (2 ft/sec) to ensure scour, but less than 1.5 m/s (5 ft/sec) under normal flow conditions to avoid excessive head loss. During peak flow conditions, the velocity should be less than 3 m/s (10 ft/sec). If these velocities are being exceeded, the force main may be too small. The pipe may have been undersized initially or more customers may have come online recently, thereby exceeding the capacity of the pipe.

Another problem is that the individual pumps may be too large for the force main, as illustrated in the following example. A 600-m (2000-ft) -long, 150-mm (6-in.) force main with a C-factor of 100 carries an average flow of 550 m^3/d (100 gpm) and an expected peak flow of 2200 m/d (400 gpm) (minor losses are negligible). First, determine the system head curve, given that the average lift is 6 m (20 ft). Second, determine the average operating point for the two different constant-speed pumps (A and B) given in Figure 4.5. Third, for each pump, determine the percentage of time and number of hours the pump will run during the year. Finally, determine the energy cost for each pump if the wire-to-water efficiency for each is 60% and the price of energy is 8 cents/kWh.

The system head curve can be determined by inserting $D = 150$ mm (6 in), $L = 610$ m (2000 ft), and $C = 100$ into the Hazen-Williams equation and adding the lift of 6 m (20 ft) to yield the following:

$$h = [10.4(2\,000)/6^{4.87}](Q/100)^{1.85} + 20 = 0.000\,675\,Q^{1.85} + 20$$

The system head curve generated from this equation is plotted on Figure 4.5.

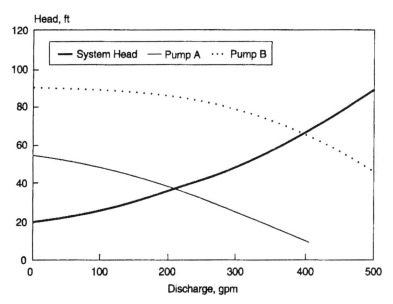

FIGURE 4.5 Pump comparison (ft × 0.304 8 = m; gpm × [6.308 × 10⁻⁵] = m³/s).

From Figure 4.5, the operating point for pump A is 1100 m³/d (200 gpm) at 12 m (38 ft) of head; for pump B, it is 2200 m³/d (400 gpm) at 20 m (64 ft) of head. Because the average flow is 550 m³/d (100 gpm), pump A will run 50% of the time, or 4380 hours per year, while pump B will run 25% of the time, or 2190 hours per year.

The annual energy cost for each pump can be determined as follows:

Cost (A) = [(2.53 × 10⁻⁴)(200 gpm)(38 ft)(0.745)(4 380 hr/yr)($0.08/kWh)]/0.6 = $837/yr

Cost (B) = [(2.53 × 10⁻⁴)(400 gpm)(64 ft)(0.745)(2 190)($0.08/kWh)]/0.6 = $1,409/yr

The aforementioned example illustrates that even though each pump was selected with the same efficiency, energy price, and total annual discharge, the pump that produced the lower friction head loss resulted in greater energy savings. This is because the velocity in the pipe was twice as high when the larger pump (B) ran. The smaller pump (A) still produced adequate velocity (0.7 m/s [2.3 ft/ sec]) when it ran. If the utility was concerned about infrequent peak flows and wanted greater peak capacity, it could achieve this by adding another parallel pump like A rather than using pump B.

5.5   Low Flows

One situation in which energy savings can be realized occurs when a pump has been sized for future growth but, in the early years of its operating life, the flows are considerably less such that the pumping station is not running most of the time. In this situation, the user can save money by shaving the impeller or replacing the impeller with a downsized impeller, and saving the original impeller for when flows increase. The smaller impeller can pay for itself in a few years provided it is not so small that loss of efficiency is excessive or that the velocities in the force main fail to reach scour velocity. In other instances, providing a small "jockey pump" to run in off-peak hours can result in energy savings.

5.6   Pipe Restrictions

In addition to a force main being undersized to carry the required flow, excessive head loss can also be a result of some type of restriction in the flow. Therefore, it is desirable to record discharge pressure from pumping stations. If that pressure should increase abruptly, then it is possible that someone has left a valve partly closed or some large object has entered the discharge pipe. An abrupt drop in suction head can result from a partly closed valve or a foreign object entering the line on the suction side of the pump.

  If the increase in discharge head or decrease in suction head occurs gradually over time, then sedimentation in the pipe, scaling, corrosion, or air blockage are more likely reasons for a rise in discharge head and/or a decrease in discharge flowrate. Checking air-release valves or installing valves at high points along the force main can correct problems with air blockage. Sedimentation or encrustation can best be corrected by cleaning the line via pigging or scraping.

  Cleaning the force main increases the carrying capacity, which moves the system head curve to the right, increases pump discharge, decreases pump head, and generally saves energy. An exception to this is, for example, if a pump has been sized to pump into a discharge line with a low carrying capacity and the carrying capacity actually turns out to be much higher, as illustrated in the following example.

  A pump is discharging into a 3.2-km (2-mile) force main with a Hazen-Williams C-factor of 80 and a diameter of 200 mm (8 in). After cleaning the pipe by pigging, the C-factor was increased to 110. The lift head is negligible compared to the head required to overcome friction losses. The pump head and wire-to-water efficiency curves are given in Figure 4.6. Plot the system head curve, determine the operating

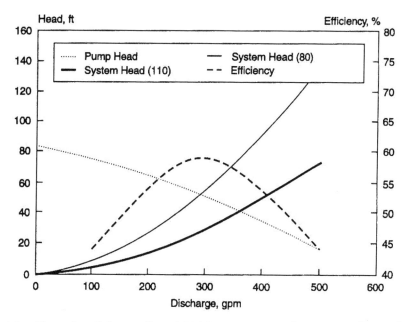

FIGURE 4.6   Changing C-factor (ft × 0.304 8 = m; gpm × [6.308 × 10⁻⁵] = m³/s).

points for both the original and cleaned line, determine the number of hours each pump must run during the year to pump an average flow of 820 m³/d (150 gpm), and determine the annual energy cost for each operating point based on an energy cost of 7 cents/kWh.

The Hazen-Williams equation is used to develop the system curve as follows:

$$h = [(10.4)(10\ 560)/8^{4.87}](Q/C)^{1.85}$$

For $C = 80$, the equation simplifies to

$$h = 0.001\ 32 \times Q^{1.85}$$

For $C = 100$, the equation is

$$h = 0.000\ 734 \times Q^{1.85}$$

The system head curves are plotted in Figure 4.6. They show that for $C = 80$, the operating point corresponds to a flow of 1635 m³/d (300 gpm), a head of 15 m (50 ft), and a wire-to-water efficiency of 60%. For $C = 110$, the operating point corresponds to a flow of 1962 m³/d (360 gpm), a head of 12 m (40 ft), and an efficiency of 55%.

Because the average flow during the course of a year is 817 m^3/d (150 gpm), the number of hours each pump must run can be determined as follows:

$$t = (150/300)(8\ 760) = 4380 \text{ hours for } C = 80$$
$$t = (150/360)(8\ 760) = 3650 \text{ hours for } C = 110$$

The annual energy cost can be calculated as

$$\text{Annual Cost} = [(2.53 \times 10^{-4})(300)(50)(0.745)(4\ 380)(7/100)]/0.6 = \$1{,}445/\text{yr}$$
$$\text{Annual Cost} = [(2.53 \times 10^{-4})(360)(40)(0.745)(3\ 650)(7/100)]/0.55 = \$1{,}261/\text{yr}$$

The aforementioned example shows that the annual energy savings was $184/year by pipe cleaning. The reason the savings were not greater is that the efficiency of the pump dropped because it was sized to pump against the larger head. Conversely, if the pump were at a maximum efficiency of 60% at $C = 110$ (Annual Cost = $1,159/yr) and up from an efficiency of 55% when $C = 80$ (Annual Cost = $1,580/yr), the annual energy savings by cleaning the pipe would have been $421/year, which is substantially greater. Therefore, it is necessary to check the actual operating points when attempting to estimate the benefits of a pipe-cleaning project.

6.0   PRICE OF ENERGY

In addition to saving energy, the ultimate goal of most utilities is to save money. When the price of energy is constant over time, the monetary savings is proportional to the energy savings. However, power companies understand that the value of energy is not constant; rather, it is much more valuable during peak energy use times. Unlike water, it is impossible to store energy for use during peak periods. Therefore, many power companies have tariffs that reflect the differing value of energy as a function of the time of day and season. Details of energy pricing are described elsewhere in this manual. However, in summary, there is a premium paid for use of energy during peak energy use times and discounts are often available for pumping during off-peak times.

In terms of wastewater pumping, there is little potential for taking advantage of time-of-day energy pricing because wastewater needs to be pumped when it arrives at the pump station. In water distribution pumping, it is possible to take advantage of time-of-day pricing by filling elevated tanks during off-peak hours and minimizing pumping energy use during peak-price hours.

The extent to which time-of-day pricing can be used depends on the volume of available storage and the level to which operators will allow tank water levels to drop during peak price times. The amount of savings can be determined by running an extended period hydraulic analysis coupled with energy cost calculations. In water systems with multiple pressure zones, it is important to fill tanks in the highest pressure zone because water delivered to those high zones has a higher energy cost and, hence, more potential for savings in using off-peak-price energy to fill the tanks. Because the cost of energy is increasing with time, the cost used in the analysis should reflect future energy trends minus inflation. This will tend to make capital investments that reduce energy use more attractive.

7.0 PUMP EFFICIENCY

As can be seen from pump curves, there is an optimal flowrate at which any centrifugal pump should operate. This is referred to as the *best efficiency point,* although there is no guarantee that a pump will operate at this point. The point at which a pump operates is determined by the intersection of the pump head curve and the system head curve. The loss of efficiency in a pump is due to turbulence, friction, and recirculation within the pump. The efficiency of a pump is not a constant, but varies with the flow through the pump as shown for some pumps in Figure 4.7.

Consider the pumps shown in Figure 4.7a with two pumps that have essentially the same pump head curve and the head curve for the system they pump into is the same. Hence, they have the same operating point. However, each has a different pump efficiency curve. When a single pump is running, the operating point is 2180 m^3/d (400 gpm) at 15.2 m (50 ft) of head. Pump A will run at an efficiency of 70%, whereas pump B will run at 62%.

However, if two pumps are running as shown in Figure 4.7b, then the station's overall operating point is 2670 m^3/d (490 gpm) at 18.9 m (62 ft) (1335 m^3/s (245 gpm) per pump). In this case, two pumps like pump A will run at an efficiency of 60%, whereas two pumps like pump B will run at an efficiency of 68%. Therefore, it is important to understand not only the operating point when a single pump is running, but also the operating point and pump efficiencies when multiple pumps are running (if multiple pumps are expected to be running frequently).

8.0 DRIVE EFFICIENCY

Variable-speed drives also introduce inefficiency in pumping. Older variable-speed drives such as eddy couplings, hydraulic couplings, primary voltage control, and slip

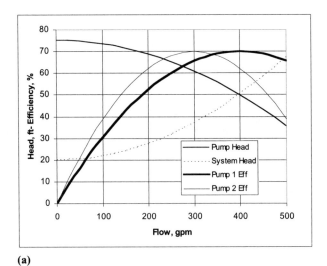

(a)

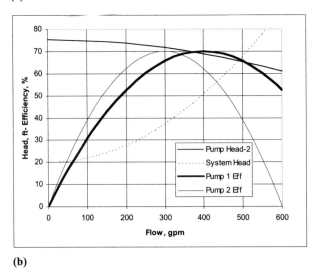

(b)

FIGURE 4.7   Pump head and efficiency characteristic curves and system head curve.

ring regenerative drives could be inefficient even at only a small reduction in speed. Modern VFDs tend to be more efficient even at a significant turndown in speed. However, even the best VFDs result in some inefficiency. One need only feel the heat coming off an operating VFD to understand this loss of efficiency.

The efficiency of a VFD depends on a number of factors such as size of the motor, load, and relative turndown (ratio of actual speed to full speed). It is difficult to find data on loss of efficiency as a function of turndown. Manufacturers often only give the efficiency at full speed, which is of little importance in evaluating VFD performance at different speeds. In general, drive efficiency can be low as the speed is turned down. Such data should be requested from the manufacturer so that a fair evaluation of variable-speed pumping can be made.

Other factors to consider with variable-speed pumping include the following (Walski, 2005):

- Variable-speed pumping may result in low velocities that can lead to solids deposition in pipes.

- Variable-speed drives add extra components to a pumping system that can fail and require maintenance; therefore, be sure to specify the ability to bypass the VFD if it should fail.

- Running pumps at other than their rated flowrate can result in bearing wear for which the pump was not designed.

- Variable-speed pumping tends to reduce hydraulic transients at startup and shutdown, but do not eliminate transients due to power failure, which are often the worst transients.

- In general, it is better to turn a pump off rather than run it at low speed.

Some calculations showing the superiority of variable-speed pumping are based on comparing a variable-speed pump with pumping through a throttled control valve (which is inefficient). For water and wastewater systems, a comparison should be made between variable-speed pumping and turning the pump off.

Variable-speed drives are discussed in more detail in Chapter 5. Additional guidelines on the use of variable-speed pumping can be found in the 2004 publication, *Variable Speed Pumping: A Guide to Successful Applications,* by the Hydraulic Institute (Parsippany, New Jersey) and Europump (Brussels, Belgium).

9.0 MOTOR EFFICIENCY

Unlike pump and drive efficiency, motor efficiency does not vary widely with flowrate or turndown unless the load drops well below the design load. Motors can be specified with a range of efficiencies including standard, high, and premium. Using a life cycle cost analysis, it is possible to determine whether it is economical to specify a high or premium efficiency motor. In most situations, when a pump will be run a significant portion of the time (i.e., it is not simply an emergency/fire backup pump), specifying a high or premium efficiency motor will be justified. (See Chapter 3 for more details on motors.)

10.0 LIFE CYCLE COSTING

Pumps and associated piping should not be selected simply based on the lowest initial cost, but rather on overall life cycle costs. There is a tradeoff between the size of the piping and the present worth of energy used to pump water. The combination of the initial cost of the pump and pipe and the present worth of the energy cost that yields the lowest cycle costs should typically be selected. Life cycle costing for pumps is described in more detail in the Hydraulic Institute's 2001 publication, *Pump Life Cycle Costs: A Guide to LCC Analysis for Pumping Systems*.

Once the annual energy cost is determined using equations given earlier, the present worth of those costs can be determined by multiplying the annual cost by the series present worth factor given by:

$$\frac{P}{A} = \frac{(1+i)^N - 1}{i(1+i)^N} \qquad (4.16)$$

Where

P/A = series present worth factor,

i = interest rate expressed as a decimal, and

N = number of years (usually 10 to 20 for pumping equipment).

When sizing a simple force main/pipeline, it is possible to perform the cost analysis with a simple spreadsheet. Figure 4.8 shows the kind of results that can be expected for a pipe carrying 2179 m³/d (1000 gpm). For small-diameter pipes, the construction cost is low but the energy costs are high. For large-diameter pipes, the energy cost is low but the construction cost is high. In this example, either the 200- or 250-mm (8- or 10-in) pipe has the best life cycle costs. For more complicated force mains or water distribution systems, hydraulic models are usually needed to determine operating points and energy costs.

11.0 OPERATION AND MAINTENANCE PRACTICES

Pumps require routine inspection to ensure that neither packing nor seals are leaking excessively, motors are not overheating, excessive vibration is not occurring, and that bearings are properly lubricated. Additionally, pumps should be inspected routinely

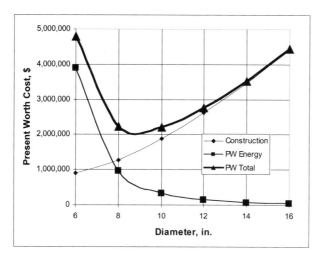

FIGURE 4.8   Determining pipe diameter based on life cycle costing.

for impeller wear and pitting, casing wear, ring wear, and misalignment. With proper care, alignment, and repair or replacement of worn and pitted parts, pumps will perform at or close to their design ratings. With well-maintained pumps, energy efficiency is strictly a matter of system design, system condition, and how the pumps are operated. In some simple systems, little can be done operationally to affect pumping efficiency, whereas, in complex systems, several things can be done to improve efficiency. Most systems fall in between.

It is important for operators to monitor supervisory control and data acquisition information. For example, if two "identical" pumps alternate between lead and lag and pump 1 discharges considerably more than pump 2, there may be a blockage, partly closed valve, or damaged impeller in pump 2. This will only be detected if operators actively review operational information and not simply wait for alarms. Proper maintenance will keep the pump at or near its original design efficiency rating and proper setting of controls can enable pumps to work more efficiently (Pincince, 1970; Yin et al., 1996).

Best management practices involving O&M practices for pumps can be found in

- *Water Distribution Operator Training: Pumps and Motors* by the American Water Works Association (AWWA) (Denver, Colorado) and

- Chapter 8 ("Pumping of Wastewater and Sludge") of Water Environment Federation's (WEF's) 2007 edition of *Operation of Municipal Wastewater Treatment Plants.*

12.0 FLOW METER CALIBRATION VERIFICATION

Field measurement of flows is difficult to perform with great accuracy. Numerous techniques are available for measuring flow, with new devices being introduced to the market each year (see U.S. EPA [1981] or Walski [1984]). Wastewater can coat sensors or plug gauges, and pipe walls can become coated with material. All of these factors can reduce the accuracy of flow-measuring devices, which means that constant maintenance and routine calibration are necessary. This is addressed further in WEF's 2006 edition of *Automation of Wastewater Treatment Facilities*.

Because *flow* is defined as the volume passing a point at a given time, the most sound way to verify the calibration of a flow meter is to measure the change in volume in some kind of vessel during a period of time. This can be done in a pumping station wet well and is referred to as a *drawdown test*. This is discussed further in AWWA's 2003 manual, *Transit Time Flow Meters in Full Closed Conduits*. To perform such a test, it is necessary to measure the plan area of the wet well (subtracting out large objects) and accurately measure the change in water level (with a staff gauge or pressure transducer). The primary source of error is inflow to the wet well during the test. If such flow is small, it can be ignored. If it is larger, the inlet can be plugged for a short period of time. If it is larger still, the inflow can be measured for a period of time with the pumps off, and an estimate of the inflow rate can be added to the pump discharge.

For example, a wet well measuring 3 m × 6 m (10 ft × 20 ft) with approximately 0.7 m^2 (8 sq ft) is taken up by equipment and columns. Inflow was estimated at 223 m^3/d (2.6 L/s or 41 gpm) with the pumps off. With one 2200-m^3/d (400-gpm) pump running, the water level dropped by 0.67 m (2.25 ft) during a 10-minute test. Find the pump discharge as follows:

$$\text{Discharge} = \text{Volume Change}/\text{Time} + \text{Inflow}$$

$$(0.7 \text{ m})[(3 \text{ m} \times 6 \text{ m}) - 0.7 \text{ m}^2] \times 1000/(10 \text{ min} \times 60 \text{ sec/min}) + 2.6 \text{ L/s}$$
$$= 20 \text{ L/s} + 2.6 \text{ L/s} = 22.6 \text{ L/s} \ (364 \text{ gpm})$$

Or

$$= 2.25 \text{ ft} \times [(10 \text{ ft} \times 20 \text{ ft}) - 8 \text{ sq ft}] \times (7.48 \text{ gal/cu ft}/10 \text{ min}) + 41$$
$$= 323 \text{ gpm} + 41 \text{ gpm} = 364 \text{ gpm}$$

Table 4.3 lists examples of best management practices for energy conservation for pumping systems.

TABLE 4.3 Examples of best management practices

Description	Comments
Need to actively look for wasted energy	A pump will not tell the operator it is wasting energy
Base pump selection on life cycle costing	Energy cost is usually the single largest cost in the life of a pump. Don't simply select a pump based on low purchase cost.
Reduce flow to be pumped	Consider conservation and leak reduction for water systems and infiltration and inflow (I/I) reduction for wastewater systems.
Maintain pumps properly	Good maintenance can prevent deterioration of efficiency. Check clearances and replace wear rings as needed. Check impellers for damage. Correct the source of cavitation problems.
Consider variable-speed pumping for systems with no tanks/wet wells	Need to evaluate based on life cycle costing. Remember that variable-speed drives are not perfectly efficient, especially when speed is turned down.
Investigate pump combinations when selecting pumps	Some pumps may run efficiently when run alone but perform badly when run in combination with other pumps
Consider high- and premium-efficiency motors	Determine life cycle cost savings to determine if justified.
Turning off a pump saves more energy than slowing down pump speed	In some systems, a pump may need to be kept running at all times. Usually, it is acceptable to allow the wet well level to fluctuate.
Field check pump efficiency	Perform periodic pump efficiency tests. At a minimum check the flow and head at the operating point to ensure it is on the original pump curve.
Analyze tradeoffs between energy cost and force main size	Larger diameter force mains have less head loss and can result in lower life cycle cost as long as velocity is adequate.
Negotiate energy tariff	Sometimes a large customer can negotiate a better energy tariff.
Calculate what the energy use/cost should be	Some computer models have the capability to determine these values easily to compare with actual energy use/cost.
Monitor pump station energy bills	Look for trends that indicate energy is being wasted.
Consider hydropneumatic tank for water systems with low flows	Energy savings can sometimes pay for the tank, especially for systems like campgrounds or resorts where flow can be essentially zero for extended periods of time.
Don't pump into pressure reducing valve	Look at the entire system to determine when flow is lifted to a higher pressure zone only to be dropped down into a lower zone.
Periodically revisit pump station energy performance	A pump may have been selected correctly but the water system may have changed (e.g., new tank added) or the wastewater system may have changed (e.g., new pump station added to a common force main). Use a hydraulic model to check how the system is performing with new upgrades. Pump may need to be changed.
Maximize pumping during off-peak hours in systems with time-of-day energy pricing	Try to store as much water (and energy) in elevated tanks during off-peak times.

13.0 REFERENCES

American Society of Civil Engineers (1992) *Pressure Pipeline Design for Water and Wastewater;* American Society of Civil Engineers, Committee on Pipeline Planning: New York.

Jones, G. M.; Sanks, R. L.; Tchobanoglous, G.; Bosserman, B. E., II, Eds. (2008) *Pumping Station Design,* 3rd ed.; Butterworth-Heinemann: Woburn, Massachusetts.

Ormsbee, L. E.; Walski, T. M. (1989) Developing System Head Curves for Water Distribution Pumping. *J. Am. Water Works Assoc.,* **81** (7), 63.

Pincince, A. B. (1970) Wet-Well Volume for Fixed Speed Pumps. *J. Water Pollut. Control Fed.,* **42** (3), 126.

Walski, T. M. (1984) *Analysis of Water Distribution Systems;* Krieger Publishing: Malabar, Florida.

Walski, T. M.; Chase, D. V.; Savic, D. A.; Grayman, W. M.; Beckwith, S.; Koelle, E. (2003) *Advanced Water Distribution Modeling and Management;* Haestad Press: Waterbury, Connecticut.

Walski, T. M. (2005) The Tortoise and The Tare. *Water Environ. Technol.,* **17** (6), 57–61.

Yin, M. T.; Andrews, J. F.; Stenstrom, M. K. (1996) Optimum Simulation and Control of Fixed-Speed Pumping Stations. *J. Environ. Eng.,* **122** (3), 205.

14.0 SUGGESTED READINGS

Hicks, T. G.; Edwards, T. W. (1971) *Pump Application Engineering;* McGraw-Hill: New York.

Hydraulic Institute (2009) *ANSI/HI Pump Standards;* CD-ROM. Hydraulic Institute: Parsippany, New Jersey.

Hydraulic Institute (1983) *Hydraulic Institute Standards for Centrifugal, Rotary & Reciprocating Pumps,* 14th ed.; Hydraulic Institute: Cleveland, Ohio.

Karassik, I.; Messina, J.; Cooper, P.; Head, C. (2008) *Pump Handbook,* 4th ed.; McGraw-Hill: New York.

Mays, L. W. (2000) *Water Distribution Systems Handbook;* McGraw-Hill: New York.

U.S. Environmental Protection Agency (1981) *NPDES Compliance Flow Measurement;* MCD-77; Washington, D.C.

University of Florida (1986) *Operations and Training Manual on Energy Efficiency in Water and Wastewater Treatment Plants;* University of Florida, Center for Training, Research and Education for Environmental Occupations: Gainesville, Florida.

Walski, T. M.; Barnard, T. E.; Merritt, L. B.; Harold, E.; Walker, N.; Whitman, B. E. (2004) *Wastewater Collection System Modeling and Design;* Haestad Press: Waterbury, Connecticut.

Water Environment Federation (1993) *Design of Wastewater and Stormwater Pumping Stations;* Manual of Practice No. FD-4; Water Environment Federation: Alexandria, Va.

Chapter 5

Variable Controls

1.0 INTRODUCTION

Several methods are used to control the output of pumps, blowers, and other powered equipment in water and wastewater facilities. This chapter discusses current technology used by the various methods as well as the pros and cons of each method. Methods range from direct control of the driver (motor or engine) to control the speed or force between the driver and the driven element, and the driven element itself. The control method itself introduces another device (with its own level of inefficiency) that must be taken into consideration. Variable or adjustable controls are used for several reasons, but they do not lead to greater system efficiency in all cases. Adjustable-speed drive (ASD) technologies are also discussed in this chapter, with a brief description of the various technologies as well as their pros and cons from an energy savings perspective in water and wastewater facilities. Finally, anticipated energy savings are not realized in some applications because some of the losses associated with the most prevalent ASD technologies in use at water/wastewater facilities, including installation and operation and maintenance, were not taken into consideration. For example, variable frequency drives (VFDs), which represent the most prevalent technology in use at these facilities, are approximately 95 to 97% efficient and motor efficiency typically begins to decrease at less than 75% of full load (see Chapter 3, "Electric Motors and Transformers," for additional details on how motors contribute to energy efficiency). In addition, the quality of electric power supplied to the motor can affect both its efficiency and its power rating (U.S. Department of Energy, Industrial Technologies Program, 2006).

Adjustable controls are often used to provide a continuous, uninterrupted output, which varies with either the demands of the input or the demands of the output as imposed by the system or the operator. In the case of a main wastewater pump at a wastewater treatment plant (WWTP), it is generally better to have a uniform flow rather than to have a single pump abruptly turning on and off, causing surges through the system. However, in larger WWTPs using multiple pumps, it may be more efficient to use constant-speed pumps for the base load, with one variable-controlled pump adjusting to the changing flow conditions. (For a detailed discussion of pump operations, refer to Chapter 4, "Pumping.") It should be noted that VFDs need to be evaluated on a case-by-case basis to determine actual savings. Such evaluations should include a life cycle cost analysis because the savings can be impacted by several system factors.

According to an energy study performed by Pacific Gas and Electric Company (San Francisco, California) (2003), a detailed analysis of each installation should be performed to determine the effectiveness of applying variable-speed drive (VSD) technologies to WWTP processes. The report states the following:

"Energy savings from VFDs can be significant: A VFD controlling a pump motor that usually runs less than full speed can substantially reduce energy consumption over a motor running at constant speed for the same period. For a 25 hp motor running 23 hours per day (2 hours at 100% speed: 8 hours at 75%: 8 hours at 67%: and 5 hours at 50%) a VFD can reduce energy use by 45%. At $0.10 per kWh, this saves $5,374 annually. Because this benefit varies, depending on system variables such as pump size, load profile, amount of static head and friction, it is important to calculate benefits for each application before specifying the use of a VFD.

Given the large diurnal flow variation in many municipal wastewater facilities, it is important to develop curves of actual flow (gallons per minute or million gallons per day) at hourly time increments during a typical day. This variable flow curve can be overlayed on the pump or blower system head/capacity curve and a baseline consumption developed assuming a single speed motor and a throttling valve to achieve the required hourly flow. A daily time weighted energy usage for the pump or blower can then be established using the pump or blower efficiency at each hourly flow rate. This baseline energy consumption can then be compared to achieving the required hourly flow rate using a variable speed drive. This analysis can be repeated for each pump or blower, and for typical summer and winter diurnal flow rates and wet weather flows.

In the design of new wastewater treatment plants or major expansions the actual selection of the type, capacity and number of pumps or blowers is complex. Flexibility to meet flow variation can be achieved by using multiple units sized for some fraction of maximum flow, with one or more units equipped with VFDs. The successful application of VFDs is also a function of the head against which the pump or blower must operate. In applications where a large static head must be overcome, VFDs may not be effective, as a very small reduction in speed can result in an excessive reduction in flow and head.

2.0   TYPES OF VARIABLE CONTROLS

2.1   Indirect Control

The output of a pump, to a small degree, self-regulates to the demands being placed on it. For example, as wet well elevation varies, the output of a centrifugal pump will vary. Its discharge will increase as intake elevation increases and its discharge will decrease as intake elevation decreases. Although this may not provide the degree of operator control desired, it must be taken into account in pumping station design.

2.2   Control of the Driver

2.2.1   *Motors*

Motors can be controlled directly via various electrical means. Direct current motors are seldom used in water treatment plants and WWTPs and, therefore, will not be discussed here. Alternating current motors are the most commonly used motors and can be controlled with electronic drives (refer to Chapter 3 for additional details on commonly used motors). Several technologies are currently available and will be discussed in this chapter. However, VFDs have been preferred in recent years because of their high efficiency, reliability, and reasonable cost. As such, VFDs will be a focus of this chapter.

2.2.2   *Engines*

Engines are controlled simply by varying the supply of fuel; the more fuel supplied, the greater the power produced. Although it is possible to use internal combustion engines as a driver for pumps in water and wastewater plants, other aspects must be considered when making such selections (e.g., emissions requirements, maintenance, and so on). The engine fuel can be gasoline, diesel fuel, digester gas, or natural gas. Depending on the rating, cooling can be liquid or air. Energy savings are possible when using digester gas in a wastewater plant. Digester gas is composed of approximately 65 to 70% methane and has a heating value of 20.5 to 26 MJ/m^3 (550 to 700 Btu/cu ft) (Karassik et al., 2008).

2.3   Motor Control

2.3.1   *Adjustable-Speed Drives or Variable-Speed Drives*

Induction motors are the most common and efficient types of motor used in water/WWTPs and other industrial applications. The most effective way to control the speed of an induction motor is to vary the frequency of its power supply based on the following formula for motor synchronous speed:

$$S = 120\,f/p \tag{5.1}$$

Where

S = motor synchronous speed (rpm),
f = frequency (Hz), and
p = number of poles in the motor.

Variable-frequency drives are designed to operate on this basic principle to take advantage of the induction motor's inherent principle of operation. They do so by varying the voltage and frequency without changing the ratio of their nominal voltage and frequency (i.e., keeping a constant volts/hertz ratio). Single-speed drives start motors abruptly, subjecting the motor to high torque and current surges up to 10 times the full-load current. In contrast, VSDs provide a "soft start" capability, gradually ramping up a motor to operating speed. This lessens mechanical and electrical stress on the motor and reduces maintenance costs and extends motor life. Variable-speed drives allow more precise control of processes such as wastewater pumping, water distribution, aeration, and chemical feed. Several different technologies exist for VSDs, however, the most common is the pulse-width-modulated (PWM) VFD type coupled to an alternating current induction motor. Variable-frequency drives work with most three-phase electric motors, so existing pumps and blowers that use throttling devices can be retrofit with these controls. Careful attention should be paid when using existing motors for retrofit applications because existing motors may not be inverter-duty rated or have the proper insulation. In addition, pump operating speeds and ambient temperatures as well as room ventilation (for indoor locations) must be considered because low-speed operation will reduce motor cooling on totally enclosed fan-cooled (TEFC) motors. Although many newer VFD designs do not require the use of inverter-duty motors, the designer should always verify motor compatibility with the VFD vendor and/or motor supplier.

2.3.2   Variable-Frequency Drives

Variable-frequency drive controllers are solid-state electronic power conversion devices. The typical design first converts alternating current input power to direct current intermediate power using a rectifier bridge. The direct current intermediate power is then converted to quasi-sinusoidal alternating current power using an inverter switching circuit. The rectifier is usually a three-phase diode bridge, but controlled rectifier circuits are also used. The circuitry on these drives typically use Insulated Gate Bipolar Transistors (IGBTs). Drives are typically characterized as either low voltage (i.e., less than 2 kV) or medium voltage (i.e., 2.0 through 6.6 kV) (see Figure 5.1 for a typical inverter system.) For a typical wastewater facility in North America, low voltage is usually considered less than 600 VAC (Volts Alternating Current), with 480 VAC a common voltage distribution; for medium-voltage distribution, a typical range is 2300 to 4160 VAC. The National Fire Protection Association's (Quincy, Massachusetts) (2008) *National Electric Code* (Article 328.1) defines medium voltage as 2 kV or higher.

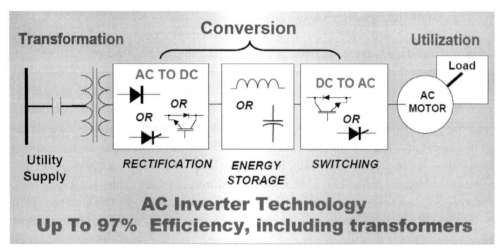

FIGURE 5.1   Typical inverter system (courtesy of Toshiba International Corporation).

2.3.2.1   Low-Voltage Drives

Low-voltage drives are commonly used for applications in the low-to-medium power range. The exact breakdown where low-voltage drives versus medium-voltage drives are used varies with user preference and can range from 373 kW (500 hp) and higher.

2.3.2.2   Medium-Voltage Drives

For large industrial loads that require multi-megawatt electric motors, one growing option is the medium-voltage alternating current drive. In addition to delivering the necessary power to an application, medium-voltage VFDs produce less power loss than low-voltage drives and allow smaller cable sizes between the motor and power source. This yields better overall drive efficiency and lower system costs. Although hundreds of times larger, medium-voltage drives are similar to low-voltage drives, which are more common. Traditionally custom-engineered medium-voltage drives now are competitive with low-voltage drives when installed costs are compared (i.e., medium-voltage drives require substantially smaller gauge wiring due to decreased current requirements). With the advent of higher voltage-rated (1.7- to 6.5-kV) power semiconductor technologies (i.e., IGBTs and symmetrical gate-commutated thyristors [SGCTs]), medium-voltage drives are becoming widely available at competitive prices because of their savings in energy costs. Specifically, substantial energy savings are possible in powering large fans, pumps, and compressors, which are typical applications for medium-voltage drives.

In many instances, the lower cost of new medium-voltage drives allows users to enjoy a payback in less than 1 year (Bartos, 2000). With the advances in power semiconductor switches, new devices such as IGBTs and SGCTs improve packaging, increase reliability, and reduce overall drive cost.

2.3.2.3 *Scalar Versus Vector Control*
There are two types of PWM inverters:

- Scalar and

- Vector.

2.3.2.3.1 Scalar. The most common type of inverter provides an output frequency that is scaled to the output voltage and is known as the volts/hertz type. This type of control provides regulation of 1 to 3% speed, setting down to 6 Hz or 10% of the maximum speed (60 Hz) by inputting the frequency setting (Karassik et al., 2008). Control algorithms are typically microprocessor-based. Alternating current drives that use PWM techniques have varying levels of performance based on the specific control algorithms used by the manufacturer. The volts/hertz control is a basic control method that provides a VFD for applications like fans and pumps. In addition, it provides fair speed and starting torque at a reasonable cost. In its simplest form, volts/hertz control takes speed reference commands from an external source and varies the voltage and frequency applied to the motor. By maintaining a constant volts/hertz ratio, the drive can control the speed of the connected motor.

2.3.2.3.2 Vector. When more accurate speed control is needed, vector drives can be used to obtain speed regulation down to 0.01% with tachometer or encoder feedback and 0.5% without feedback ("sensorless"). Vector drives represent an improvement over scalar VFDs in that they separate the calculations of magnetizing current and torque-generating current. These quantities are represented by phase vectors and are combined to produce the driving-phase vector, which in turn, is decomposed into the driving components of the output stage. These calculations are done in the drive's microprocessor. There are three basic types of vector control for alternating current drives used today: sensorless vector control, flux vector control, and field-oriented control. Sensorless vector control provides better speed regulation and the ability to produce a high starting torque. Flux vector control provides more precise speed and torque control with dynamic response. Field-oriented control drives provide the best speed and torque regulation available for alternating current motors. Vector control drives provide direct current-like performance for alternating current motors and are well suited for typical direct current applications. One of the basic principles of a flux vector drive is to simulate the torque produced by a direct current motor (full torque at zero speed). Until the advent of flux vector drives, slip had to occur (30 to 50 rpm for motor torque to be developed) where slip is the difference between synchronous speed and rotor speed (see motor chapter for additional details on motor operation). Slip also increases with motor load. The actual slip is a dimensionless number and is defined as: (Fink, Beaty 1987).

$$S = N_s - N/N_s \qquad (5.2)$$

Where

S = slip
N_s = synchronous speed
N = actual speed of motor

With flux vector control, the drive forces the motor to generate torque at zero speed (Threvatan, 2006).

Vector drives often require feedback such as a tachometer to provide shaft speed, although some manufacturers have developed sensorless vector drives that do not require this feedback. For water and wastewater pumping applications, the sensorless type of vector is sufficient because the added accuracy in speed regulation is not warranted. In water and wastewater facility applications, the sensorless vector is the most prevalent drive. It is used for diverse applications, such as metering pumps, as well as for the following:

- Dosing chemicals,

- Return activated sludge pumps,

- Waste activated sludge pumps,

- Heating, ventilating, and air conditioning (HVAC) fans and hot water recirculating pumps,

- Sludge dewatering and thickening centrifuges speed control of bowl and scroll to maintain speed differential between bowl and scroll. One of the main advantages of a sensorless flux vector over standard PWM is higher starting torque on demand (Threvatan, 2006).

It is also worth noting that microprocessor technology is rapidly making scalar drives obsolete except in the smallest of sizes. The low cost for processing power has made the issue of having and maintaining separate designs untenable for many manufacturers.

Vector drives essentially use a mathematical model of the motor that, together with current vectors, make it possible to determine and control the actual motor speed. Current vectors are made up of magnetizing (horizontal) and torque-producing (vertical) vectors. To continually control the current, one must compensate for the angle of voltage displacement (power factor) and the time constant (magnetic and thermal delay) of the rotor. These are dynamic conditions that vary with different load demands and temperatures. In essence, the vector control decouples the excitation current from the torque-producing current, similar to a direct current drive where the field excitation current is separated from the armature current. More detailed descriptions of vector drive controls can be found in *Variable Speed Drive Fundamentals* (Phillips, 1999).

2.3.2.3.3 Motor Compatibility. To account for the added winding heat created by harmonics, motors are typically derated 5 to 10% when used with VFDs. Initially, standard alternating current motors were used on inverter drives. Most motor manufacturers offer "inverter-duty" motors specifically geared to VFD applications. Inverter-duty motors provide improved performance and reliability when used in variable-frequency applications. These special motors have insulation designed to withstand the steep-wave-front voltage impressed by the VFD waveform, and are redesigned to run smoother and cooler on inverter power supplies.

Many new motors have a Class F (155 °C) insulation system whereas older motors may only have a Class B (130 °C). Most inverter-duty motors have a Class H system rated for 180 °C operation. As stated earlier, many drive manufacturers are now offering products that do not require the use of inverter-duty motors. In addition, most new motors meeting the National Electrical Manufacturers Association's (Rosslyn, Virginia) (2007) *Specification for Motors and Generators* comply with the inverter-duty motor requirements. (Refer to Chapter 3 for additional information on types of motors and their various applications for energy savings.)

In some applications, particularly those using 4-pole or 6-pole motors, additional short-term-capacity increases for pumps and blowers may be achieved by "over-speeding" the motors (i.e., operating at greater than 60-Hz output frequency). In applications where the motor is oversized to accommodate design discharge pressure or head significantly greater than normal operating head, this may result in energy savings by allowing one pump or blower to operate instead of two units at greatly reduced capacity. Because horsepower requirements on a centrifugal pump increase by the cube of the speed change, it does not take much more speed to overload the motor and risk premature failure; as such, careful consideration should be given to the application when choosing this option.

2.3.3   Other Technologies

There are other technologies used for speed control of motors in water and wastewater plants (Karassik et al., 2008). However, because they do not offer energy savings comparable to VFDs they will not be discussed here in any length. Such technologies include (all of which are used with wound-rotor induction motors):

- Liquid rheostats,
- Step resistors, and
- Slip-loss recovery.

This section is included because there are still several facilities which use the above types of control for their main wastewater pump's speed control applications. However, they are no longer installed because they cannot compete with VFD technology. Indeed, once VFD technology became reliable and popular, wound-rotor systems and these technologies were no longer a choice of designers and industry, including water/wastewater utilities.

The speed of a wound-rotor motor is varied by removing power from the rotor windings. This is usually accomplished by switching resistance into the rotor circuit via the use of slip-rings and brushes. As resistance is added, the speed of the motor decreases. However, this method is not efficient because of the loss in the resistors. In addition, additional cooling is required to remove all the excess heat from the drive cabinet. Starting torques of the wound-rotor induction motor can be varied from a fraction of rated full-load torque to break down torque by proper selection of the external resistance value. The motor is capable of producing rated full-load torque at standstill with rated full-load current. The motor has low starting current, high starting torque, and smooth acceleration. In general, however, speed stability below 50% of rated full speed is unsatisfactory. Additional maintenance is required because of the slip rings and brushes required to access the rotor windings. Another speed control method for pumps not widely used is variable speed fluid drives (eddy current couplings, permanent magnet). See Table 5.1 for a comparison of different speed control technologies.

TABLE 5.1 Comparison of various speed control methods

Method	Advantages	Disadvantages
Fixed-speed squirrel cage induction motors with modulating control valve	Uses standard induction motors and starters	• Adds additional complexity to process control, additional components to maintain (valve, controller, actuator) • Requires control valve with actuator (pneumatic or electric) • Energy inefficient • Places more stress on equipment downstream
Two-speed induction motors	Two-speed motors and starter are readily available and simple to control	• Two speeds do not provide operational flexibility for range of variable pumping rates required by process
Wound-rotor motor with step resistors	Step resistors and associated controls are simple devices requiring minimal skills to maintain	• Requires custom solutions • Multiple steps may be insufficient to satisfy process demands for variable pumping • Technology is outdated with limited suppliers • Speed switching contactors have limited life • Energy inefficient; requires additional HVAC cooling • Requires more floor space than a comparable VFD • Motors not suitable for classified areas • Limited availability of motors and controls • Motors are more expensive • Motors are not as rugged and reliable as standard induction motors with squirrel cage rotors. Carbon brushes and slip rings require more maintenance
Liquid rheostat	Simple controls	• Obsolete technology, limited suppliers • Requires wound-rotor motor with associated maintenance and procurement issues • Limited speed range, generally 2 to 1 • Poor efficiency (85% at full speed to 45% at half speed) • Poor power factor at low speed • Requires a heat exchanger with circulating pump to dissipate from the liquid in the rheostat the heat produced by the electrical slip in the motor to obtain variable-speed operation • Not suitable for constant torque loads such as positive-displacement type pumps
Magnetic coupling/ eddy current coupling	• Wide speed range (34 to 1) • Uses constant speed motor • High torque applications	• Low efficiency at reduced speed • Cooling required by air or liquid for magnetic coupling at low speed; water cooling required above 400 hp • Problems in maintaining proper alignment between motor, magnetic coupling, and pump • Additional floor space for horizontal drives and head clearance for vertical drives required

2.3.4   Harmonics

2.3.4.1   Description

Harmonics are voltages and currents in the electrical system at frequencies that are multiples of the fundamental frequency (i.e., 50 Hz in European power systems and 60 Hz in the United States and other countries) (Europump et al., 2004). Harmonic waveforms are characterized by their amplitude and harmonic numbers. Variable-frequency drives draw current from the line only when the line voltage is greater than the direct current bus voltage inside the drive. This occurs only near the peaks of the sine wave. As a result, all of the current is drawn in short intervals (i.e., at higher frequencies). Variation in VFD design affects the harmonics produced. For example, VFDs equipped with direct current link inductors produce different levels of harmonics than similar VFDs without direct current link inductors. Variable-frequency drives with active front ends using transistors in the rectifier section have much lower harmonic levels than VFDs using diodes or silicon-controlled rectifiers (SCRs).

2.3.4.2   Other Sources of Harmonics in Plants

Electronic lighting ballasts, uninterruptible power, supplies, computers, office equipment, ozone generators, and other high-intensity lighting are also sources of harmonics. Plant harmonic analysis must include these sources when calculations are performed for VFDs.

Harmonics are the result of non-sinusoidal current, which is characteristic of all adjustable-frequency controls using power semiconductors such as diodes, IGBTs, and SCRs on the input (National Electrical Manufacturers Association, 2001). Harmonics are thus present in waveforms that are not perfect sine waves due to distortions from nonlinear loads such as VFDs (see Figure 5.2).

The harmonics of primary concern from a PWM VFD are the 5th, 7th, 11th, and so on. The frequencies of these harmonics are 300 Hz, 420 Hz, 660 Hz, and so on. These low-order harmonics typically are the largest magnitudes present on the line. Harmonics that are multiples of 2 are not harmful because they cancel out. The same is true for third-order harmonics (third, sixth, ninth, and so on). Because the power supply is three-phase, the third-order harmonics cancel each other out in each phase. This leaves only the 5th, 7th, 11th, 13th, and so on to discuss. The magnitude of the harmonics produced by a VFD is greatest for the lower order harmonics (5th, 7th and 11th).

The Institute of Electrical and Electronics Engineers (IEEE; New York, New York) (1993) publishes a standard, *Recommended Practices and Requirements for Harmonic Control in Electric Power Systems* (IEEE 519-1992), that defines harmonic limits on

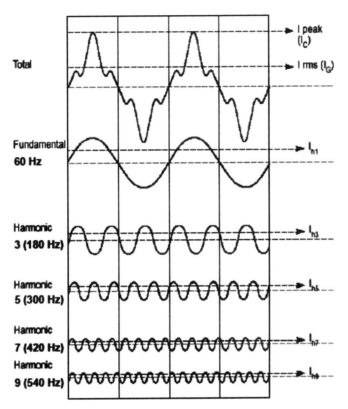

Figure 5.2 Harmonic waveforms (courtesy of Schneider Electric).

power systems. In Europe, harmonic limits are set under the Electromagnetic Compatibility (EMC) directive (Europump et al., 2004). Harmonic distortion levels, as stated in the IEEE 519-1992 standard, are intended to be applied at the point of common coupling (PCC) between the utility system and other users. This standard was created to protect the utility's customers from harmonics generated by facilities that use harmonic-generating equipment (National Electrical Manufacturers Association, 2001). However, most water and wastewater plants as well as other industrial facilities use the standard as a guide to maintain harmonic levels inside the plant in compliance with limits imposed by the standard. Furthermore, IEEE 519-1992 defines limits at the PCC because the power company pays for the infrastructure up to the PCC. The user bears the cost of the distribution system within his or her own facility as well as any oversizing that may be required.

$$\text{THD} = \sqrt{\frac{\text{sum of all squares of amplitude of all harmonic voltages}}{\text{square of the amplitude of the fundamental voltage}}} \cdot 100\%$$

$$\text{THD} = \frac{\sqrt{\sum_{h=2}^{50} V^2 h}}{V_1} \cdot 100\%$$

FIGURE 5.3   Voltage THD calculation as per IEEE 519 (courtesy of © IEEE Standard 519 [1993]).

Harmonic distortion measurements are typically given in "total harmonic distortion" (THD). Per IEEE 519-1992, Section 8.6, THD is used to define the effect of harmonics on the power system voltage. Total harmonic distortion is used in low-voltage, medium-voltage, and high-voltage systems; it is expressed as a percent of the fundamental and is defined as shown in Figure 5.3.

Total Harmonic Distortion (THD) defines the harmonic distortion in terms of the fundamental current drawn by a load. Per IEEE 519-1992, Table 11-1, THD limits are 5% on voltage at the PCC for voltages of 69 kV or less. Total harmonic distortion for current is not defined in the standard.

Current distortion limits are dependent on total plant load and individual equipment load. These are defined in IEEE-519, Section 10.4. The IEEE 519-1992 standard uses a term called *Total Demand Distortion* (TDD) to express current distortion in terms of the maximum fundamental current that the consumer draws. The limits IEEE 519-1992 places on current distortion also depend on the ratio of I_{sc}/I_{load}, where I_{sc} is the short-circuit current and I_{load} is the load current drawn by the VFD. The short-circuit current is determined by the plant's electrical short-circuit study and is typically based on the plant's utility supply transformer. The short-circuit current for a supply transformer can usually be obtained from the utility. The ratio of I_{sc}/I_{load} determines the "stiffness" of the supply. Therefore, the "stiffer" the supply, the higher the ratio of I_{sc}/I_{load} will be and the more current TDD allowed.

2.3.4.3 Effect on Efficiency

Harmonics increase equipment energy losses and may create excessive currents and heating in transformers and neutral conductors (Europump et al., 2004). As a result, harmonics contribute to lowered quality and efficiency and reduced system capacity, and require costly equipment upgrades to support expansion. Additional effects of harmonic distortion include:

- Overheating and sustained damage to bearings, laminations, and winding insulation on generators, causing early life failure; there are similar effects on transformers: in the United States and Japan, for instance, harmonics have resulted in fires in distribution transformers.

- Overheating and destruction of power factor correction capacitors; there is a danger of resonance (i.e., voltage amplification) if the "detuning reactors" are not fitted to the capacitor bank, causing catastrophic damage to the capacitor bank and other equipment.

- Overheating of the stator and rotor of fixed-speed electric motors; risk of bearing collapse due to hot rotors. This is especially problematic on explosion-proof motors with increased risk of explosion; if the voltage distortion is over the prescribed limit stated on the certification, the motor is no longer certified, losing any third-party assurance as to its safety.

- Spurious tripping of electrical circuit breakers. Interference with electrical, electronic, and control system equipment, including computers, radio communications, measuring devices, lighting, and so on.

- Overheating of cables and additional risk of failure due to resonance; a decreased ability to carry rated current due to "skin effect," which reduces a cable's effective cross-sectional area.

Harmonics can also have detrimental effects on a plant's electrical system. The harmonics created by the VFD can be picked up by other electrical lines that have common connections with the VFD. Therefore, care must be exercised when specifying VFDs for energy efficiency applications. A harmonic analysis of the entire electrical distribution system should be done to determine any effects of harmonics on the system. Such analysis is typically performed with commercially available software and can be done by a VFD manufacturer as part of the purchase price of the VFD equipment or by an independent party who can perform a plant-wide analysis of the entire electrical distribution system (U.S. Department of Energy, Industrial Technologies Program, 2006).

2.3.4.4 Harmonics Mitigation Methods
Typical mitigation methods include the following:

- *Passive Filters*—This is an economical solution for drives below 149 kW (200 hp). May cause a power system resonance condition in some installations. Use caution when applying this type of filter with generator power sources (see Figure 5.4).

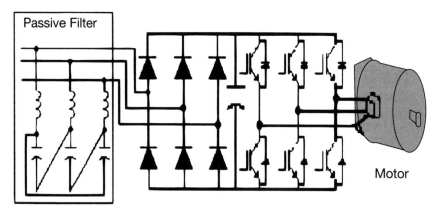

FIGURE 5.4 Passive filter (courtesy of Rockwell Automation, Inc.).

- *Active Filters*—This is an external solution that actively monitors harmonic distortion levels and injects cancellation harmonic current onto the line to meet IEEE 519-1992 at the power line where the active filter is connected. Cost-effective only on large common alternating current bus applications where several drives are connected on the bus. This solution can also be applied to existing installations where there are harmonics present on the bus.

- *Input Line Reactors*—Adding this reactor reduces transient surges or spikes on the line, but provides little added harmonic mitigation. It is sometimes used on distribution systems composed of a small percentage of nonlinear load as compared to total load.

- *Active Front End*—This solution actively tracks and regulates input current to maintain sine wave current draw. This technique generates minimal voltage distortion, allowing the input power converter to meet IEEE 519-1992 requirements at the input terminals of the drive. This method is most cost-effective on large common bus systems with many drives. For medium-voltage drive applications, this is typically the most cost-effective solution (see Figure 5.5).

- *Multipulse Rectifiers with Phase-Shifting Transformers*—This solution includes 12 and 18 pulse solutions, which rely on two or three separate three-phase systems, each feeding a diode or silicon controlled rectifiers (SCR) bridge. The direct current output is then combined to feed the capacitor in the direct current bus. Each of the three-phase input sections is phase shifted from the other

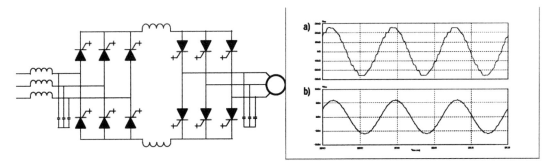

FIGURE 5.5   Pulse-width-modulated rectifier (active front-end) and its input current/voltage waveforms, (a) line current and (b) line-to-line voltage at PCC (courtesy of Rockwell Automation, Inc.).

by 60 deg/n, where "n" is the number of three-phase feeds. Thus, an 18-pulse system requiring three separate three-phase feeds would have a phase shift of (60 deg/3) 20 deg to supply three sets of full-wave bridge rectifiers. This type of system is effective if all of the three-phase feeders have balanced voltage. It also requires one rectifier section for each three-phase feed and a special transformer to produce the multiple secondary-phase shifted outputs. The 18-pulse PWM VFDs meet the general requirements of IEEE 519-1992 at the drive input terminals without analysis; they also require the use of phase-shifting transformers at the drive inlet. Theoretically, this configuration cancels the 5th, 7th, 11th, 13th, 23rd, 25th, 29th, 31st harmonics, and so on. The resulting system draws a pulse of current every 20 electrical degrees and typically produces harmonics at 4% THD or less. Most manufacturers currently use autotransformers instead of separate winding transformers for this purpose in order to conserve space and improve system efficiency. Separate winding transformers add to system cost by requiring additional space, cooling, and wiring for a complete installation (see Figure 5.6).

Newer solutions include 24-pulse designs. Figure 5.7 contains a circuit diagram of a 4160-VAC 24-pulse design. Input alternating current is supplied through an input controller to transformer *T1*. The transformer has four isolated secondary windings per output phase, each feeding a three-phase full-wave rectifier bridge. The output of the rectifiers is connected to three inverter power modules that produce three-phase alternating current power at the frequency and voltage required by the motor.

Converter Pulses	Harmonic Order								THD
	5	7	11	13	17	19	23	25	Up to 49th
6	0.175	0.110	0.045	0.029	0.015	0.009	0.009	0.008	27.0%
12	0.026	0.016	0.073	0.057	0.002	0.001	0.020	0.016	11.0%
18	0.021	0.011	0.007	0.005	0.045	0.039	0.005	0.003	6.6%
AFE	0.037	0.005	0.001	0.019	0.022	0.015	0.004	0.003	5.0%

FIGURE 5.6   Harmonic reduction using multipulse converters (courtesy of Rockwell Automation, Inc.).

- *Generator Operation*—An important consideration when using VFDs in water and wastewater plants is the operation of VFDs on emergency generators. The harmonics study must take this into account because compliance under emergency generator operation is harder to attain due to the high impedance characteristics of most generators. In addition, it is recommended that generators be oversized and not loaded with nonlinear loads such as VFD loads in excess of 20% of generator capacity. When passive filters are used on generator power, a filter should be selected with a dropout contactor terminal block for the filter capacitors to limit the leading power factor at no-load operation.

2.3.5   Motor-Bearing Damage

Another potential problem with the use of VFDs, particularly medium-voltage VFDs, is that they may cause motor-bearing fluting or damage caused by shaft voltages if the proper protective steps are not taken. However, if the VFD/motor system is properly designed, these problems are rare and can be avoided. Solutions include:

- Insulated bearings, or shaft grounding brushes, to minimize arcing in the bearing;
- Low PWM carrier frequency;
- Output filter to reduce the rate of rise of voltage of each pulse and reduce the capacitive-induced voltage in the shaft; and
- Shielded cables between the VFD and the motor.

2.3.6   Common-Mode Noise

Common-mode noise is a type of electrical noise induced on signals with respect to ground. Faster output instantaneous rate of voltage charge (dv/dt) transitions of IGBT drives increase the possibility for increased common-mode electrical noise. If not addressed, this can affect other plant equipment. Variable frequency drives must

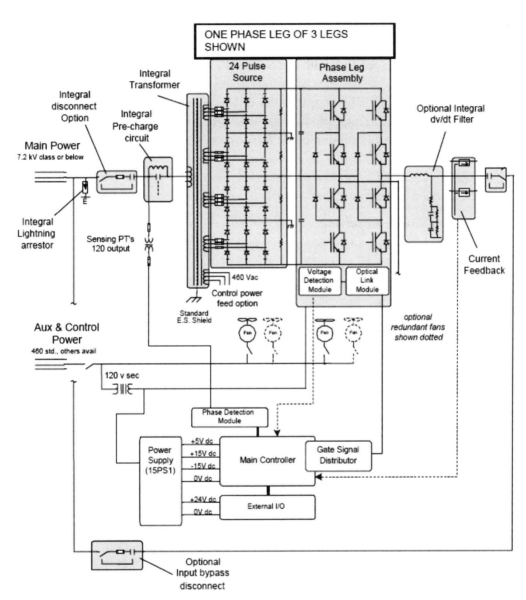

FIGURE 5.7 Twenty-four-pulse medium-voltage VFD (courtesy of Toshiba International Corporation).

be grounded properly to avoid common mode noise, intermittent operation, and nuisance tripping of the drive. All drives must be grounded per manufacturer's recommendations and in compliance with the National Electric Code (NEC) requirements. Grounding wire from the motor must return to the VFD via separate ground wire(s) in situations where the VFD is grounded to the plant ground. Another effective method of reducing common-mode noise is to attenuate it before it can reach the ground grid. Installing a common-mode ferrite core on the output cables can reduce the amplitude of the noise to a level that makes it relatively harmless to sensitive equipment or circuits. Common mode cores are most effective when multiple drives are located in a relatively small area.

2.3.7   Speed Control Considerations for Pumps, Blowers, and Compressors

Manufacturers of VFDs offer two types of drives: variable torque and constant torque. The two are basically the same, except that the constant torque drive has a slightly larger kilovolt-ampere rating. The selection depends on the load being driven. Variable-torque applications are the best candidates for adding an ASD to save energy. Centrifugal pumps and fans represent variable-torque loads, where the amount of power required drops off by the cube of the speed decrease. Thus, the actual savings come from reducing the motor speed, which results in lowering the motor's power requirements.

Constant torque applications such as conveyors often use ASDs, although this is typically not for energy savings, rather, for production flow improvements. Based on affinity laws (see also Section 4.1 later in this chapter), if the pump speed is reduced to 50% of the rated speed, the required torque is reduced to 25% of the rated torque. The torque produced by the motor that drives the pump or fan is proportional to the product of the air gap flux and the stator winding current. The VFD that powers the motor keeps the air gap flux constant at all speeds (constant volts/hertz ratio). Therefore, the motor current must also be reduced to 25% of the rated current. The motor I2R, where I is the motor stator current and R is the stator winding resistance, losses reduce to 6.25% of the losses at the rated current. Motor and VFD would have no problem dissipating the heat. (Note: Typical I2R losses for a motor running full speed typically account for 20 to 30% of total motor losses.)

For constant torque loads such as rotary and screw conveyors, elevator drives, hoists and cranes and reciprocating pumps, motor torque—which must be the same as the load torque during steady-state operation—is the same at all speeds. Motor efficiency is lower and heat dissipation is a problem because of the reduced air flow from the shaft-mounted fan. This problem is more severe in TEFC motors used in Class 1, Division 2, hazardous locations, particularly when the motor operates at reduced speed for extended periods of time.

3.0 BLOWERS AND COMPRESSORS

See Chapter 9 "Blowers" for a complete discussion of blower controls and optimization.

Multistage centrifugal blowers have limited turndown capability (typically 70%) and lower efficiencies than single-stage units. Single-stage blowers with variable inlet vanes and variable discharge diffusers allow flow adjustments while maintaining constant impeller speed. They are capable of compression efficiencies of up to 80% from full output down to 40% output. Disadvantages are higher cost and noise levels. Centrifugal blower head capacity curves are flat, and the discharge pressure is reduced by the square of the blower speed, so a small change in speed can cause such a reduction in pressure as to be unable to overcome the static head of the aeration basin. Variable-frequency drives for application to aeration blowers must have very precise control of the variable speed. The output of rotary lobe, positive-displacement blowers cannot be throttled. However, capacity variation can be obtained by using multiple units or VSDs.

4.0 OPTIMIZATION OF PUMP OPERATION

4.1 Affinity Laws

Centrifugal pumps and fans are dynamic devices that generate head by a rotating impeller and follow the affinity laws. The affinity laws are equations relating the pump performance parameters of flowrate, head, and power absorbed, to speed as follows (Europump et al., 2004):

$$Q \propto n \tag{5.3}$$

$$H \propto n^2 \tag{5.4}$$

$$P \propto n^3 \tag{5.5}$$

Where

Q = flowrate;
H = head;
P = power absorbed;
n = rotational speed

For many systems, VFDs can help to improve pump operating efficiency despite changes in operating conditions. The effect of slowing pump speed on pump operation is illustrated by the three curves in Figure 5.8.

When a VFD slows a pump, its head/flow and brake horsepower curves drop down and to the left, and its efficiency curve shifts to the left. This efficiency response provides an essential cost advantage; keeping the operating efficiency as high as possible across variations in the system's flow demand can reduce the energy and maintenance costs of the pump significantly. Variable-frequency drives can also be used with positive-displacement pumps (U.S. Department of Energy, Industrial Technologies Program, 2006).

To illustrate savings that are possible, Phillips (1999) calculated the energy cost for constant-speed versus variable-speed fans. Although the figures provided represent 1999 dollars, they illustrate the types of savings possible using variable-speed pumping. For more recent cost calculations, software tools are available like the U.S. Department of Energy's (U.S. DOE's) Industrial Technologies Program Pumping System Assessment Tool, which provides estimates of optimal efficiency (U.S. Department of Energy, Industrial Technologies Program, 2006). The Pumping System Assessment Tool and its user manual can be downloaded from U.S. DOE's best practices Web site at http://www1.eere.energy.gov/industry/bestpractices/software.html#psat.

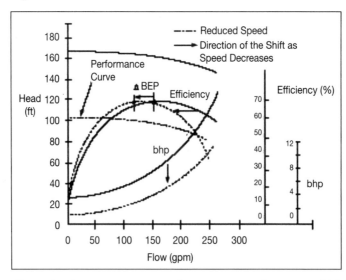

FIGURE 5.8   Effect of reducing pump speed on pump performance (U.S. DOE, 2006).

4.2 Drive Energy Savings Measurements

A simple way to determine actual energy savings in drives is reading the kilowatt meter function that is available on the drive keypad. Compare the keypad reading to the calculated kilowatts that the motor would normally consume if operated at full speed and at rated frequency:

$$kW = \text{Volts} \times \text{Amps} \times 1.732 \times PF \qquad (5.6)$$

Where

   PF = the power factor.

The drive uses a mathematical algorithm to determine the kilowatt reading. Therefore, to be sure the reading is accurate, check the drive programming for the proper settings for the actual (measured) incoming system voltage and the motor nameplate data of horsepower, revolutions per minute (RPM), voltage, and full load amps.

5.0 REFERENCES

Bartos, F. (2000) Medium Voltage AC Drives Shed Custom Image. *Control Eng.* [Online], Feb 1, 2000; http://www.controleng.com/article/CA191379.html?industry=Discrete+Control&industryid=22073&spacedesc=communityFeatures&q=Medium+Voltage+AC+Drives+Shed+Custom+Image (accessed Dec 2008).

Carlson, R. (2000) The Correct Method of Calculating Energy Savings to Justify Adjustable-Frequency Drives on Pumps. *IEEE Transactions on Industry Applications,* Vol. 36, No. 6, November/December.

Europump, Hydraulic Institute, and U.S. Department of Energy, Industrial Technologies Program (2004) *Variable Speed Pumping, A Guide to Successful Application;* www1.eere.energy.gov/industry/bestpractices/pdfs/variable_speed_pumping.pdf (accessed Jan 2009).Evans, I. (2002) Harmonic Mitigation for AC Variable Frequency Pump Drives. *World Pumps* [Online], Dec 2002; http://www.worldpumps.com (accessed Jan 2009)

Fink, D. G; Beaty, W. H. (1987) p 20–28, Standard Handbook for Electrical Engineers, 12th Edition, McGraw-Hill Inc.

Institute of Electrical and Electronics Engineers (1992); IEEE-std 519-1992 *Recommended Practices and Requirements for Harmonic Control in Electric Power Systems;* IEEE Press: New York.

Karassik, I.; Messina, J.; Cooper, P.; Heald, C. (2008) *Pump Handbook,* 4th ed.; McGraw-Hill: New York.

National Electrical Manufacturers Association (2001) *Application Guide for AC Adjustable Speed Drive Systems;* NEMA Standard Publication; National Electrical Manufacturers Association: Rosslyn, Virginia.

National Electrical Manufacturers Association (2007) *Specification for Motors and Generators;* MG-1, Section IV, Part 31 (Rev. 1 2007); National Electrical Manufacturers Association: Rosslyn, Virginia.

National Fire Protection Association (2008) *National Electrical Code;* NFPA-70; National Fire Protection Association: Quincy, Massachusetts.

Pacific Gas and Electric Company (2003) *Municipal Water Treatment Plant Baseline Energy Study;* PG&E New Construction Energy Management Program; SBW Consulting: Bellevue, Washington.

Peeran, S. M. (2008) VFDs and Motors: Making the Right Match. *Consult.-Specifying Eng.* [Online], July 1, 2008; http://www.csemag.com/article/CA6578954.html (accessed Jan 2009).

Phillips, C. A. (1999) *Variable Speed Drive Fundamentals,* 3rd ed.; Fairmont Press: Lilburn, Georgia.

Pump Systems Matter and Hydraulic Institute (2008) *Optimizing Pump Systems: A Guide for Improved Energy Efficiency, Reliability, & Profitability;* Pump Systems Matter: Parsippany, New Jersey:

Threvatan, V. (2006) *A Guide to the Automation Body of Knowledge,* 2nd ed.; ISA Press: Research Triangle Park, North Carolina.

U.S. Department of Energy, Industrial Technologies Program (2006) *Improving Pumping System Performance: A Sourcebook for Industry,* 2nd ed.; U.S. Department of Energy, Office of Energy Efficiency and Renewable Energy: Washington, D.C.

U.S. Department of Energy, Industrial Technologies Program (2000) *Energy Management for Motor Driven Systems;* Rev. 2.; U.S. Department of Energy, Office of Energy Efficiency and Renewable Energy: Washington, D.C.

Weber, W.J.; Cuzner, R.M.; Ruckstadter, E.J.; Smith, J. (2002) Engineering Fundamentals of Multi-MW Variable Frequency Drives—How They Work, Basic Types, and Application Considerations; *Proceedings of the 31st Turbomachinery Symposium;* Houston, Texas, Sep. 9-12; Texas A&M University: College Station, Texas.

Chapter 6

Energy Use in Water Facilities

1.0 GENERAL OVERVIEW

Energy consumption in water facilities will vary significantly based on the location and topography of the service area, plant size, treatment processes used, and age and condition of the water transmission system. Transporting water—either raw water from the source to the treatment plant, treated water from the plant to the distribution system, or in-plant pumping—consumes the majority of energy in water facilities. Advanced water treatment processes such as ozonation, high-pressure membrane filtration, and/or desalination and UV light disinfection also account for a significant portion of the power consumption associated with water treatment. These new emerging technologies that result from a combination of having to treat more unconventional water sources, such as seawater or highly saline groundwater with increasingly stringent regulations to provide a "multiple-barrier approach," also require high-quality uninterruptable power supply and advanced sensitive control systems, thus requiring additional provisions for backup power supplies as part of risk management. A benchmarking study by the Water Research Foundation (formerly the American Water Works Association Research Foundation [AwwaRF]) (Denver, Colorado) concluded that energy costs account anywhere between 2 to 35% of utility operating budgets (Jacobs et al., 2003). Currently, there are some 60 000 public community water utilities in the United States (Electric Power Research Institute, 2000). The Electric Power Research Institute (EPRI) (Palo, Alto, California) has been conducting energy audits at water and wastewater facilities for over a decade and estimates that almost 3 to 4% of the nation's electricity (200 billion to 270 billion MJ [megajoules]/a [56 billion to 75 billion kWh/yr]) is used for treating and distributing potable water and treating wastewater (Electric Power Research Institute, 2000). Almost 80% of the total energy is used for transporting water. Based on numerous audits conducted by EPRI, the average electricity consumption by a 10 mgd (40 ML/d) surface water facility is estimated to be 14 057 kWh/d or 1337 J/L (1406 kWh/mil gal). The EPRI study estimates the national average for conventional surface water plants is in the range of 670 to 1700 J/L (700 to 1800 kWh/mil. gal), depending on use and customer sector.

For example, Figure 6.1 shows the specific energy use for the Tampa Bay Water's Surface Water Treatment Plant (Tampa, Florida). The treatment plant is rated for a daily treatment capacity of 66 mgd (250 ML/d), and treatment is comprised of enhanced coagulation using a ballasted flocculation process followed by lime addition, ozone disinfection, a second lime addition, biologically active granular activated

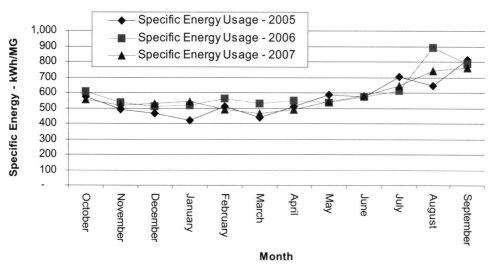

Tampa Bay Surface Water Treatment Plant - Specific Energy Usage

FIGURE 6.1 Example of specific energy use for a surface water treatment plant (data provided by Veolia Water North America and Tampa Bay Water).

carbon (GAC) filters, storage and pumping, and, finally, secondary disinfection with chloramines. The specific energy consumption (energy per unit volume of water delivered) for this plant ranges from 400 kWh/mil gal (380 J/L) in the winter to 900 kWh/mil gal (860 J/L) in the summer.

For a similar size facility treating groundwater, the power consumption is estimated to be 1735 J/L (1824 kWh/mil. gal), which is about 30% greater than that for a surface water source (Electric Power Research Institute, 1996). The difference is in the additional pumping required to transport water from deep aquifers to the surface for treatment. In certain parts of the United States, like the Chino basin in Southern California, groundwater pumping accounts for 2915 kWh/mil. gal (2772 J/L) due to the much deeper aquifers than other parts of the nation (California Energy Commission, 2005). System-wide, the three significant factors documented to increase unit energy consumption for water utilities are as follows (Electric Power Research Institute, 2002):

- Age of the water delivery system—as systems age, friction in piping increases and/or leaks may develop, resulting in higher pumping energies.

- Energy-intensive treatment processes required by ever increasing water demands and the depletion of high-quality water sources, coupled with regulatory requirements for implementing advanced treatment alternatives.

- Water conservation programs that, contrary to common belief, may actually increase unit electricity consumption as economies of scales are lost or the systems are operated at less than optimum levels. Whether water efficiency is the main focus as opposed to energy efficiency can also appear to defeat the purpose of water conservation programs. For example, California's Westland Water District has found that the most energy-efficient irrigation strategies do not provide sufficient water efficiency to achieve state targets for reduced water use. The most water-efficient irrigation strategy, the drip/microspray technology, uses 0.344 MJ/m^3 (118 kWh/ac-ft). A somewhat greater efficiency can be achieved by combining sprinkler and furrow technologies that would use 0.207 MJ/m^3 (71 kWh/ac-ft).

A recent AwwaRF study estimates that a typical urban water supply scheme consumes around 95 J/L (100 kWh/mil gal) for conveyance, 240 J/L (250 kWh/mil gal) for treatment, and about 1090 J/L (1150 kWh/mil. gal) for distribution (California Energy Commission, 2005). Energy consumption distribution by unit process at a typical conventional surface water facility is presented in Figure 6.2 (EPRI 1999). As illustrated in this figure, almost 90% of the energy consumed in a water facility is used to pump water.

An energy audit is the best procedure available to an energy-consuming facility for identifying energy conservation opportunities. The 1994 EPRI report provides the steps to be taken by a utility to perform an energy audit to identify the major energy-consuming processes in a water plant and thereby identifies energy conservation measures. Many energy audits have concluded that energy efficiency opportunities can be achieved by making process modifications. The Electric Power Research Institute has found that more energy savings are achieved by improving operating and process systems than by conventional approaches such as improving motor efficiencies. Many plants have achieved reductions in power requirements of 10 to 30% through concerted efforts to institute energy efficiency programs.

Based on the energy audits they conducted, EPRI also learned that there is a lack of standardization in the approach to developing techniques to reduce energy consumption by public water utilities. In 1999, in collaboration with AwwaRF and a few large public water utilities in the United States, EPRI developed a generic

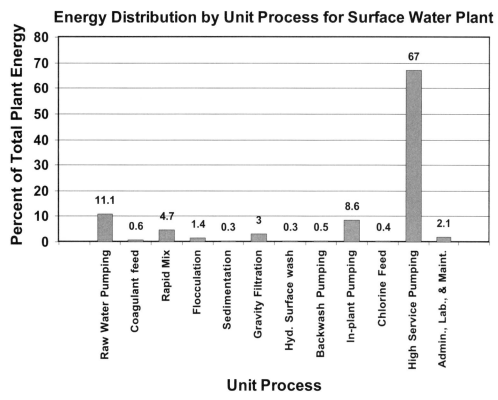

FIGURE 6.2   Energy distribution by unit process (Electric Power Research Institute, 1994).

energy and water quality management system model (Electric Power Research Institute, 1994). Development of this generic model would draw on experiences from electric utilities that have developed a similar approach to demand fore-casting for scheduling and optimizing the operation of power systems. Such a model would then be customized by the water utility to fit its own needs and, together with the utility's supervisory control and data acquisition, aid in optimiza-tion of all operations to achieve the most efficient and least energy-consuming operation for the given condition. A similar approach taken by the City of Toledo, Ohio, enabled the city to save 10 to 15% of their total energy costs in addition to reducing chemical and labor costs (Trivedi et al., 1995). The city implemented a dis-tributed control system comprising several remote microprocessors, remote ter-minal units, and a programmable logic controller (PLC). This control system uses information provided by online instruments to make operating decisions that save energy by using the most efficient combination of pumps and pump speeds to maintain proper pressures and flowrates during off-peak and on-peak energy rates.

Additional factors to consider are the geographical location and associated climate because energy costs associated with heating and ventilation of buildings and indoor facilities required by colder climates (northern United states) can be a significant percentage (80%) of the total facility energy costs (this is especially true for conventional treatment plants where most unit processes are flow by gravity) as opposed to constructing facilities out in the open with few processes being enclosed in the building (southern United States) (American Water Works Association Research Foundation, 2008).

Like EPRI, other agencies such as AwwaRF, the California Energy Commission (CEC) (Sacramento, California), the New York State Energy Research and Development Authority (NYSERDA) (Albany, New York), the Energy Center of Wisconsin (ECW) (Madison, Wisconsin), Pacific Gas and Electric Company (PG&E) (San Francisco, California), and others have conducted independent and collaborative research into energy consumption in the public water/wastewater industry and published a wealth of information on energy saving techniques that utilities across the globe could implement. The U.S. Environmental Protection Agency (U.S. EPA) has published an energy management guidebook for water and wastewater utilities that provides a road map to identify, implement, measure, and improve energy efficiency and renewable opportunities at water and wastewater utilities (U.S. EPA, 2008).

Several energy conservation case studies performed by CEC together with the California Public Utilities Commission (San Francisco, California) offer the following energy conservation techniques for water utilities:

1. Use energy-efficient motors.
2. Design pumps so that they operate in their most efficient zone. When centrifugal pumps operate at their maximum efficiency, or best efficiency point (BEP), the radial loads on the pump bearings are at a minimum because the unbalanced radial load on the impeller is at a minimum. These radial loads increase as the pump operating point moves away from the BEP, either toward shutoff or runout. Under these conditions, vibration and hydraulic losses occur within the pump and decrease the efficiency of the pump, thus affecting power consumption. For maximum operating performance, it is desirable to operate close to the BEP and to limit the range of operation between 60 and 120% of the BEP.

3. Repair and replace aging water distribution piping to reduce leaks as well as reduce pipe friction.

4. Use variable-speed pumping in lieu of throttling valves to control flow.

5. Use automated meter reading technology to detect system leaks more quickly and efficiently.

6. Replace rate-of-flow-control (ROFC) valves with hydroelectric generators. In 2003, the Southern Nevada Water Authority (SNWA) (Las Vegas, Nevada) initiated a program to replace three ROFC valves with hydroelectric generators (capacity of the generator was between 0.5 to 3 MW). The SNWA estimates its system can support a total of 10 such hydroelectric generators (20 MW total capacity).

7. Pump during off-peak hours to fill all elevated storage tanks. Use PLCs to perform such operations based on signals from remote-sensing devices installed within the system.

8. If possible, design and construct the water facility to minimize pumping. An example of this is the Denver Foothills Water Facility that is located at an elevation such that raw water is diverted from the South Platte River some 8 km (5 miles) upstream using gravity flow via a tunnel to the water plant, and treated water is again conveyed by gravity some 29 km (18 miles) to the Denver metropolitan area (Layne and Eckenberg, 1983).

9. Use backup or standby generators during hours of peak energy rates to offset energy costs.

10. Use efficient fixtures such as high-pressure sodium, metal halide, and fluorescent lamps with high-efficiency solid-state ballast.

11. Implement water conservation programs within the utility. For example, the City of Leamington, Ontario, Canada, was able to flatten its overall water demand curve by negotiating large end users to meter water 24 hours a day seven days a week and storing the water in reservoirs during low-use times for use during high-use times, thereby shifting the electricity load from peak-rate times to low-rate times.

12. Conduct routine energy audits to identify the major energy-consuming processes/equipment or facilities that could assist in optimizing the use.

The following two sections describe energy use in common conventional and advanced water treatment processes used in public water supply plants.

2.0 RAW WATER INTAKES

Raw water sources for a water plant can be either surface water such as a river, lake, or reservoir or groundwater. The typical raw water intake structure for surface water supply consists of intake gates, screens, and pumps to transfer the water from the source to the water facilities. Screens can be either manual or mechanical. Mechanical traveling automatic backwash screens are driven by small motors ((3.7 kW [(5 hp]). Intake gates can also be either manual or automatic and are driven by fractional power motors (0.7 kW [(1 hp]). Raw water intake alone is an insignificant energy-consuming process in the overall scheme of raw water treatment. However, pumping associated with raw water intake is a significant source of energy consumption and is discussed in the following section.

3.0 RAW WATER PUMPING AND CONVEYANCE

Depending on the proximity of the water plant from the raw water source and the topography of the area, raw water pumping can account for a large percentage of total energy consumed in a water facility. An example of this is the State of California, where almost 70% of the state's total stream runoff is north of Sacramento, whereas the majority of the state's water demand (almost close to 80%) is south of Sacramento. This requires pumping and conveying water several hundred miles. Additionally, the topography is such that water needs to be conveyed over the Tehachapi Mountains (about 900-m [3000-ft] high), requiring even more energy to transport water. There is a significant difference in the energy use for water conveyance between Northern and Southern California, with Southern California water utilities consuming an astounding 8500 J/L (8900 kWh/mil gal) compared to 140 J/L (150 kWh/mil. gal) for Northern California (California Energy Commission, 2005).

4.0 PRETREATMENT: COAGULATION, FLOCCULATION, AND SEDIMENTATION

This section discusses the coagulation process with respect to its energy consumption. Conventional coagulation comprises a "flash-mix" process followed by a flocculation process. The flocculated water is then provided quiescent conditions in a sedimentation tank to remove the flocculated particles from water.

4.1 Flash Mix

The purpose of flash mixing is to rapidly and completely mix the primary coagulant with the incoming process stream. Flash-mix energy is provided either by in-pipe pump-diffusion, hydraulic jet nozzles, an induction mixer, or with a vertical turbine mixer in a square concrete tank. The intensity of mixing or agitation required can be estimated using the following equation:

$$G = \left(\frac{P}{\mu V} \right)^{1/2}$$
(6.1)

Where:

G = root-mean-square velocity gradient, s^{-1};
P = power imparted to water, N m/s (ft-lb/s);
μ = dynamic viscosity, Pa s or N s/m^2 (lb-s/sq ft); and
V = process tank volume, m^3 (cu ft).

Typical G values for mechanical flash mixing range from 600 s^{-1} to 1000 s^{-1}, with typical detention times ranging from 10 to 60 seconds. For water at a temperature of 10 °C, a flowrate of 1 mgd, assuming a gear/motor efficiency of 80% for the mixing equipment, results in energy consumption in the range of 6 to 103 kWh/d, using the following equation:

Power consumption (kWh/mgd) = $G^2 \times Q \times t \times \mu \times 24 \times 0.001355 \div f$ (6.2)

The power consumption (kilowatt-hours/day) for a flash-mix process for a 1-mgd flow at a water temperature of 10 °C is provided in Table 6.1.

TABLE 6.1   Power consumption (kWh/d) estimates for flash-mix process

Power consumption (kWh/d)

Detention time - t (sec)	G value (sec^{-1})				
	600	700	800	900	1000
10	6	8	11	14	17
20	12	17	22	28	34
30	19	25	33	42	52
40	25	34	44	56	69
50	31	42	55	70	86
60	37	51	66	84	103

Assumptions: For water temperature of 10°C,
Dynamic viscosity, μ = 2.74e-05 lb.sec/ft^2
Motor/gearbox efficiency factor, f = 0.8
Conversion of ft-lb/sec to hp = 0.001818
Conversion of hp to kW = 0.7455
Motor run time, hr/day = 24

In-pipe mixing such as pump-diffusion or induction mixers requires high G values, typically 1000 s^{-1}. However, certain pretreatment chemicals such as sulfuric acid, sodium hydroxide, and antiscalants that are readily soluble in water can be thoroughly mixed using turbulence in the pipe as well as static mixers, thereby reducing energy consumption. Static mixers are typically designed to provide a maximum head loss in the pipe of 14 to 34 kPa (2 to 5 psi).

4.2 Flocculation

The function of flocculation is to provide sufficient mixing energy to agglomerate the coagulated particles into a floc large enough or dense enough to settle by gravity. Typical G values for mechanical flocculation range from 20 to 70 s^{-1}, with detention times from 10 to 30 minutes. The energy consumed in this process ranges from 3 to 114 kWh/d at 10 °C. Power consumption for a flocculation process for a 3.785-ML/d (1-mgd) flow at 10 °C is provided in Table 6.2 using the aforementioned equations and assumptions.

4.3 Sedimentation

The function of sedimentation is to provide sufficient detention time to settle large flocs formed in the flocculation basin. The only energy consumption by the sedimentation processes is for the residuals withdrawal mechanisms including pumps to transfer the settled solids to the residuals handling processes. The power required for residuals withdrawal is insignificant with respect to other processes.

TABLE 6.2 Power consumption (kWh/d) estimates for flocculation process

Power consumption (kWh/d)

Detention time - t (sec)	G value (sec^{-1})					
	20	30	40	50	60	70
600	0.4	0.9	1.7	2.6	3.7	5.1
900	0.6	1.4	2.5	3.9	5.6	7.6
1200	0.8	1.9	3.3	5.2	7.4	10.1
1500	1.0	2.3	4.1	6.5	9.3	12.7
1800	1.2	2.8	5.0	7.8	11.2	15.2

4.4 High-Rate Clarification

Fundamentally, the high-rate clarification (HRC) process is similar to conventional (coagulation, flocculation, and sedimentation) water treatment technology. Both processes require chemical coagulation to destabilize colloidal particles followed by flocculation for aggregation of the particles to increase their settling velocities. The primary advance made in the HRC process is the addition of micro-sand, typically 100 to 150 μm with a specific gravity of 2.65, as a "seed" or "ballast" to induce and promote the formation of high-density robust floc with high settling velocities. Raw untreated water is pumped into the coagulation tank of the sand-ballasted system (see Figure 6.3) where a coagulant, such as alum, ferric chloride, ferric sulfate, or poly-aluminum chloride, is added to destabilize the suspended solids and colloidal matter in the influent stream. High-rate clarification operates at significantly higher overflow rates than conventional gravity-settling processes, about 0.7 L/m² · s (1 gpm/sq ft) versus 14 to 41 L/m² · s (20 to 60 gpm/sq ft), and hence have an

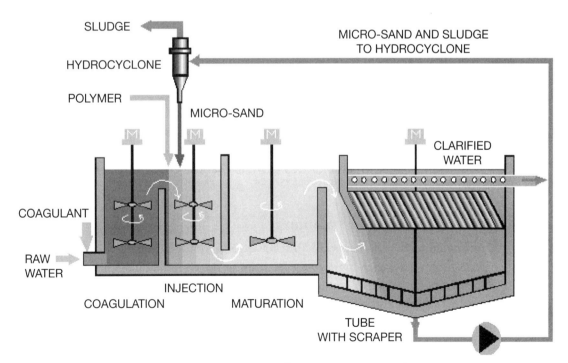

FIGURE 6.3 Schematic diagram of the ACTIFLO® process (courtesy of Kruger USA).

TABLE 6.3   Energy consumption estimates for two full-scale facilities using HRC

Process	Value	Power consumption (kWh/MG) [kWh/kiloliter]
City of Lincolnton, North Carolina		
Treatment capacity	9 mgd	
No. of treatment trains	1	
Coagulation tank mechanical mixer	1, 5 hp	9.9 [0.0026]
Injection tank mechanical mixer	1, 5 hp	9.9 [0.0026]
Maturation tank mechanical mixer	1, 7.5 hp	14.8 0.0039]
Settling tank rake mechanism	1, 1.5 hp	3 0.00079]
Tampa Bay Surface WTP, Florida		
Treatment capacity	66 mgd	
No. of treatment trains	2	
Chemical induction mixer	1, 5 hp	1.4 [0.00036]
Coagulation tank mechanical mixers	2, 25 hp	13.5 [0.0035]
Injection tank mechanical mixers	2, 25 hp	13.5 [0.0035]
Maturation tank mechanical mixers	2, 30 hp	16.3 [0.0043]
Settling tank rake mechanisms	2, 5 hp	3 [0.00079]

extremely small footprint compared to conventional coagulation/clarification units. Energy consumption for HRC is similar to that in conventional processes. Table 6.3 shows energy consumption for two facilities that use HRC technology to treat their surface waters.

4.5   Dissolved Air Floatation

Compared to Europe, especially the United Kingdom, dissolved air flotation (DAF) was seldom used in potable water facilities in the United States until a decade ago. The first potable water plant using DAF was built in 1965 in Windhoek, Namibia, for the floatation of algae-laden water. Dissolved air flotation is an excellent treatment solution for water having high levels of algae and other low-density particles that cannot be removed effectively with sedimentation. Dissolved air flotation is suitable for many other applications, including low turbidity (<30 Nephelometric Turbidity Units (NTU) water, cold water, water with high color and total organic carbon, filter backwash applications, and membrane pretreatment. A pressurized recycle stream of clarified effluent that is supersaturated with air is introduced uniformly at the bottom of the DAF tank via nozzles or orifices. The pressure drop across the nozzles releases the entrapped air from the recycle stream to create microbubbles that help with flocculation of the particles in the water. Typically, the recycle rate is 5 to 10% of the influent water flow and a compressor pumps air (at 410 to 620 kPa [60 to 90 psi] pressure) into the recycle water via an eductor to create an air/water mixture. Energy consumption depends on the size of the plant and the DAF recycle rate, and hence pump size. Based on a 5 to 10% recycle rate at 410 to 620 kPa (60 to 90 psi), energy consumption is in the range of 34 to 127 J/L (36 to 134 kWh/mil. gal).

5.0   TASTE AND ODOR CONTROL

Complaints about the taste and smell of drinking water are common for many utility systems. However, by using certain treatment techniques, plant operators may be able to reduce customer taste and odor complaints. Tastes and odors are commonly caused by the following: (1) a smell like rotten eggs caused by hydrogen sulfide (H_2S) (although H_2S does not typically pose a health risk at low levels [1 to 2 mg/L], it is still a nuisance); (2) earthy or fishy smells and tastes that can usually be traced back to algae growth in the source water; and (3) trihalomethanes, haloacetic acids, and chloramines formed by the reaction of chlorine with ammonia and organic compounds. The most common treatment methods for the elimination of taste- and odor-causing compounds are air stripping, oxidation with ozone, and adsorption with GAC. Granular activated carbon adsorption is discussed in Section 6.0, titled "Filtration," in this chapter.

5.1   Air Stripping

Air Stripping has been used in water treatment for centuries and represents the most economical process for removing volatile organic compounds and other taste- and odor-contributing compounds such as hydrogen sulfide. The principle of air stripping is to transfer the contaminant compound from the water phase to the air phase by mixing it with a sufficient quantity of fresh air for a sufficient exposure time. The transfer mechanism is governed by Henry's law for sparingly soluble gases. The higher the Henry's constant for the contaminant, the easier it is to strip the contaminant from the liquid to the gas phase, and hence the lower the required air-to-water ratio.

Air stripping can be accomplished by various means, with some as simple as diffused aeration and tray aeration or with some more complicated processes such as packed tower counter-current degasifiers. The simplest and least energy-consuming method is to use tray aerators where water trickles over a series of staked trays without any blowers. Sometimes blowers are used to force air through the trays from the bottom to improve ventilation and thus stripping efficiency. Diffused aeration consists of blowing air through a set of diffusers installed at the bottom of a process tank. Packed tower counter-current degasification involves a media that is packed in a vertical tower and where the liquid is sprayed from the top and a blower blows air from the bottom. The biggest difference in the systems is that diffused aeration provides a much larger liquid-volume-to-air interface, whereas packed tower degasifiers provide a much greater air-to-liquid volume for mass transfer. Hence, ideally, diffused aeration is typically used when the objective is to maximize dissolution of a certain gas such as oxygen or ozone in the water, whereas packed tower degasifiers should be used when the objective is to maximize the removal of a volatile compound from water to air.

The four key components in the design of packed tower degasifiers are (1) the air-to-water ratio; (2) the water velocity and thus tower diameter; (3) the type of packing; and (4) the depth of packing (Kavanaugh and Trussell, 1980). Of these four components, the air-to-water ratio and the type of packing govern energy consumption in this process. The greater the air-to-water ratio the greater the size of the aeration equipment, such as air blowers, and the greater the energy consumption. Similarly, the smaller the packing material, the larger the area available for the liquid-to-air mass transfer; however, simultaneously, the greater the pressure drop through packing media and, hence, the greater the energy consumption. Table 6.4 presents typical air-to-water ratios for various volatile contaminants found in water supplies. Henry's constants are a function of temperature and, for hydrocarbons, the constant increases about threefold for every 10 °C temperature rise (Kavanaugh and Trussell, 1980). Hence, temperature has a significant effect on the required air-to-water ratio.

Commercially available degasifier towers can handle up to 18 925 L/min (5000 gpm) of liquid and can be constructed with diameters up to 15 ft, providing a liquid application rate of 17 to 20 L/m² · s (25 to 30 gpm/sq ft). For hydrogen sulfide stripping (>90% removal), 70 000 m³/h (40 000 cfm) of air will be required for a 19 000-L/min (5000-gpm) unit. For such an application, 55.9- to 70-kW (75- to 100-hp) blowers (fans) are typically needed to provide the required air-to-water ratio. This

TABLE 6.4   Air-to-water ratios for packed tower degassifiers for various volatile organic contaminants (Stripping factor R = 3 i.e. removal of > 90%)[10] (Reprinted from *Journal AWWA*, Vol. 72, No. 12 (December 1980), by permission. Copyright © 1980, American Water Works Association.)

Volatile Organic Contaminant	Henry's Law Constant (atm) @ 20°C	Air-to-water ratio (theoretical)
Vinyl Chloride	3.55×10^5	0.011
Methane	3.8×10^4	0.11
Carbon Dioxide	1.51×10^3	2.6
Carbon Tetrarchloride	1.29×10^3	3.1
Tetrachloroethylene	1.1×10^3	3.6
Trichloroethylene	550	7.2
Hydrogen Sulfide	515	7.7
1,1,1 - Trichloroethane	400	9.9
Chloroform	170	23
1,2- Dichloroethane	61	65
1,1,2 - Trichloroethane	43	92
Bromoform	35	110
Ammonia	0.76	5200

equates to a specific energy of 167 to 240 J/L (185 to 250 kWh/mil gal) without any further air emission treatment, such as by scrubbers. Depending on the contaminant to be removed, the air-to-water ratio and the size of the equipment and energy consumption will vary appropriately. Pilot studies should be conducted for packed towers before detail design to determine the Henry's constant for contaminants at low solute concentrations over a range of temperature and mass transfer coefficients (K_{La}) for various types of packing media. The Henry's constant for other taste and odor compounds such as geosmin, 2-methylisoborneol, 2-isobutyl-3-methoxy-pyrazine, and 2-isoproyl-3-methoxy-pyrazine indicate insignificant "strippability" at neutral pH. Therefore, air stripping is economically unfeasible for removal of these compounds (Lalezary et al., 1984).

5.2   Ozone

Ozone in drinking water treatment is primarily used for color, taste, and odor removal and as a primary disinfectant. Depending on the raw water quality, ozone is added either upstream or downstream of the coagulation/flocculation process. Ozone is generated by applying a single-phase high-voltage alternating current across a small (0.3- to 3-mm) "discharge gap" filled with either air or oxygen gas. The discharge gap has glass or ceramic dielectric on one side and a stainless steel electrode on the other side. The stiochiometric specific energy required to produce ozone from oxygen is 2.952 MJ/kg 0.372 kWh/lb O_3. However, in reality, the specific energy required for medium-frequency ozone generators using liquid oxygen (LOX) as the source gas is 10 times this amount (Rackness, 2005). Conversely, ozone generators using air as the source gas require 20 times the stiochiometric specific energy to produce a pound of ozone (Rackness, 2005). There are three types of ozone generators that are typically used by municipal facilities:

- Medium-frequency, air-fed ozone generators (producing ozone at concentrations of 1 to 3%, by weight);

- Medium-frequency, oxygen-fed ozone generators (producing ozone at concentrations of 3 to 6%, by weight); and

- Medium-frequency, oxygen-fed, high-efficiency ozone generators (producing ozone at concentrations of 6 to 12%, by weight).

Figure 6.4 shows that from an energy use standpoint, medium-frequency, oxygen-fed generators require less energy to produce the same quantity of ozone than the medium-frequency, air-fed generators and the medium-frequency, high-efficiency, oxygen-fed generators. Conversely, the medium-frequency, oxygen-fed generators consume proportionately more oxygen gas to produce the same quantity of ozone compared to high-efficiency generators. Similarly, Figure 6.5 shows the increase in the quantity of cooling water required with decreasing cooling water temperature to produce the same amount of ozone. Most systems are designed around once-through, open-loop systems with a side-stream pump using finished water; depending on the ambient temperature of the cooling water, the quantity varies seasonally.

Figure 6.6 shows the consumption of LOX as a function of ozone concentration for medium-frequency generators. Thus, the cost of purchasing LOX is offset by the lower energy consumption of generating ozone. However, if the cost of LOX is much greater than the cost of energy or if the ozone concentration is varied to the effect that more LOX is used during times of peak electricity rates, the medium-frequency generators would prove highly cost effective over time.

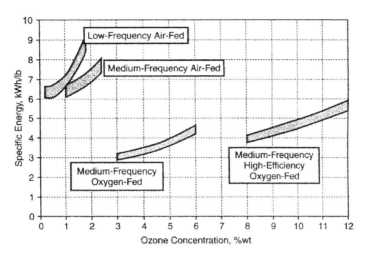

FIGURE 6.4   Specific energy profile for air-fed and oxygen-fed ozone generators (Reprinted from *Ozone in Drinking Water Treatment—Process Design, Operation, and Optimization,* by permission. Copyright © 2005, American Water Works Association.)

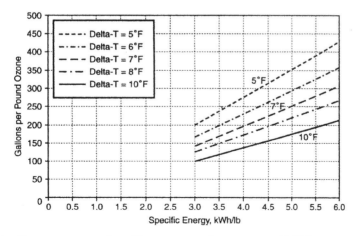

FIGURE 6.5   Generator vessel cooling water volume per unit (lb) ozone production for variable generator specific energy value, assuming power supply unit (PSU) power loss is 6% (Reprinted from *Ozone in Drinking Water Treatment—Process Design, Operation, and Optimization,* by permission. Copyright © 2005, American Water Works Association.)

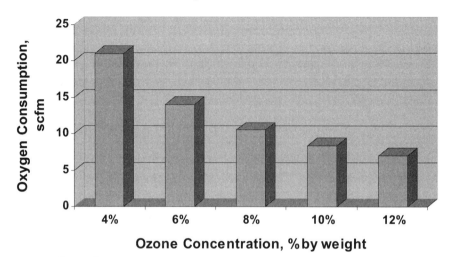

FIGURE 6.6   Liquid oxygen consumption versus ozone concentration to produce 100 lb/day of ozone for medium-frequency generators (courtesy of ITT-WEDECO).

Ozone generator power demand is a function of ozone production and the concentration of ozone and generator cooling water temperature. Figure 6.7 shows an increase in specific energy with an increase in ozone concentration. Similarly, Figure 6.8 shows a decrease in ozone production with an increase in cooling water temperature.

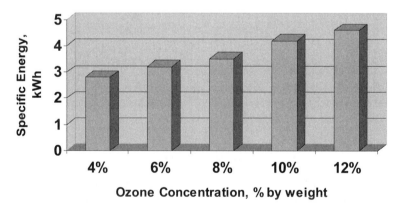

FIGURE 6.7   Power consumption versus ozone concentration to produce 1 lb/day of ozone at 60°F (16°C) cooling water for medium-frequency generators (courtesy of ITT-WEDECO).

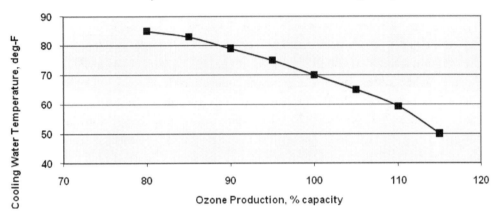

FIGURE 6.8   Effect of cooling water temperature on ozone production capacity
(courtesy of ITT-WEDECO).

Ozone system suppliers may sometimes recommend a separate chilled water
system at higher than normal ambient water temperatures. However, chillers have
not been shown to reduce overall system power usage. The quantity of cooling water
has a marginal effect on generator ozone production. The typical design cooling
water rise across the generator vessel is usually between −14 and −12 °C (7.5 and
10 °F). If the cooling water temperature is high, it is common to design with a lower
temperature rise across the vessel (closer to the −14 °C [7.5 °F] number). The approx-
imate blow per unit of energy varies based on ozone concentration in percent weight
and water temperature. For an ozone concentration of 7%, by weight, flow per unit
of energy varies between 180 and 300 L/kWh (0.8 and 1.3 gpm/kW); for 10%, by
weight, it varies between 213 and 430 L/kWh (0.94 and 1.9 gpm/kW) and, for 12%,
by weight, it varies between 250 and 495 L/kwh (1.1 and 2.18 gpm/kW) for a temperature range of 5 and 35 °C (41 and 95 °F).

Facilities have a choice of using a once-through cooling water loop (usually
nonchlorinated finished water) or an open-loop/closed-loop cooling water system
with a heat exchanger to transfer heat from the closed-loop side with the open-loop
side. Hence, facilities would need to perform a detailed operating cost comparison
using the cost of oxygen gas and the cost of power to determine which technology
would better suit their needs.

A 2003 study conducted by ECW at the University of Wisconsin, Madison that evaluated energy use at Wisconsin drinking water facilities estimated that ozone disinfection increased overall energy consumption by 120 to 550 kWh/mil gal (114 to 523 J/L) (Energy Center of Wisconsin, 2003). Similarly, an ozone facility optimization study conducted by AwwaRF in collaboration with EPRI suggests ozone energy consumption in the range of 100 to 400 kWh/mil gal (95 to 380 J/L) or more depending on water quality, system design, and process operation (Rakness and DeMers, 1998).

The Southern Nevada Water Authority has two large direct-filtration water plants (Alfred Merrit Smith, 2270 ML/d [600 mgd], and River Mountains Water Treatment Plant, 570 ML/d [150 mgd]) that use ozone as a primary disinfectant (for 2-log cryptosporidium inactivation credit). Ozone is produced at an 8% concentration, by weight, using high-purity oxygen produced onsite using vacuum/pressure swing adsorption (VPSA) equipment. A liquid oxygen facility serves as a standby for the VPSA system. Because transporting water (by pumps) represents a significant portion of SNWA's energy costs, SNWA negotiated a variable energy rate structure to take advantage of off-peak rate hours to do most of the pumping and treatment. Hence, modifications were made to the facility during construction that was originally designed to operate on a steady-state flow. (Variability in ozone demand/decay was found to be minimal for the two plants; consequently, however, plant flow would have a 6:1 turndown depending on the time of the day to save energy costs). The LOX cost in Las Vegas was found to be significantly higher than in other places in the country. Hence, optimization of LOX use was key to providing savings in operating costs. An ozone control philosophy was developed wherein an optimum cost curve would be automatically generated based on operator input for a particular time of the day/year. This would provide a setpoint ozone concentration that would then aid in selecting the number of VPSA units required and whether or not supplemental LOX is required (the energy cost will be too high above the optimum point and the oxygen gas cost will be too high below the optimum point). To automatically respond to plant flow changes, a feed-forward control strategy (for oxygen gas flow and generator power) was implemented that would facilitate quick changes in generation equipment to respond accordingly and minimize operator input. This overall strategy has reportedly saved significant operating costs for the city (Rackness et al., 2000).

Alternatively, facilities that only use LOX can implement a similar strategy based on the information generated during startup (a specific energy curve for the generator can be developed, similar to Figure 6.4). Figure 6.4 illustrates the power consumption based on operation at various ozone concentrations. Because LOX is purchased at a fixed price for an agreed-upon contract period, utilities can save daily operating costs by operating the generator at a lesser ozone concentration during on-peak energy rates. Figure 6.9 shows operating costs associated with ozone production as a function of ozone concentration. Depending on the cost of LOX and power, this graph will change. Utilities using LOX should determine their optimum point by performing a quick analysis similar to the graph as shown.

Effective transfer of ozone into water is important because the cost of producing ozone is significant especially if the ozone is carried within oxygen gas. The rate of ozone transfer and the subsequent rate of ozone decomposition depend on contact system efficiency and reaction rates of ozone with constituents in water. The ozone reaction rate is a function of water temperature, pH, and organic and inorganic constituents of water. Performing an ozone optimization study such as determining the "performance ratio" (disinfectant concentration × contact time (CT)/residual concentration × time) and unit volume operating cost (cost per megaliter or per million gallon) is highly recommended and can reduce operating costs for energy and gas. The performance ratio provides an indication of how close the measured versus the target disinfection credit is at a facility. A performance ratio less than 1 suggests required performance has not been achieved (ozone dose must be increased) and vice

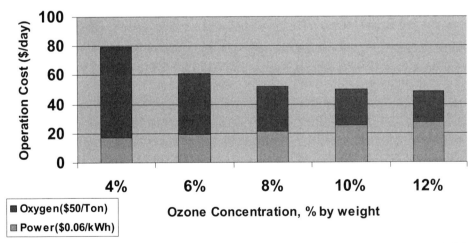

FIGURE 6.9   Daily operating cost versus ozone concentration to produce 100 lb/day of ozone at 60°F (16°C) cooling water for medium-frequency generators (courtesy of ITT-WEDECO).

versa. The Canal Road Water Treatment Plant in Somerset, New Jersey, was able to lower its operating cost by 40% by performing a performance ratio study and fine-tuning their process (Rakness and DeMers, 1998). A routine instrument calibration/standardization schedule should be developed and rigorously followed to ensure that readings are accurate. At a minimum, key instruments would be ozone residual analyzers, gas flow meters, gas-phase ozone concentration analyzers, and generator power meters.

Ozone generation alone comprises 90% of the total energy demand in an ozonation process, with the remaining 10% required for ozone generator cooling and ozone destruction processes. Some facilities (e.g., Orlando Utilities Commission's, Orlando Florida-Navy Water Treatment Plant, the City of Valdosta, Georgia, Water Treatment Plant, and many others) use air dissolution to reduce the dissolved oxygen in the ozone contactor effluent to near-saturation levels to improve water stability and prevent any corrosion in the distribution system (ozone effluent is typically supersaturated with oxygen, with concentrations in the 20-mg/L range). This requires the ozone off-gas destruct blowers to be oversized. Under such circumstances, variable-frequency drives are recommended on the off-gas destruct blowers to minimize energy consumption.

6.0   FILTRATION

6.1   Gravity Filtration

Conventional gravity filtration is the most common filter type in use in water facilities. Because gravity is the driving force for filtration, the main area for energy savings is in the filter cleaning. Filter cleaning (backwash) is accomplished by pumping clean water up through the filter. Energy savings result from running filters longer between cleanings and minimizing pumping requirements. The method of backwash dramatically affects filter run lengths due to enabling clean beds after backwash and by reducing bed stratification and subsequent surface blinding. Precise monitoring of the pressure drop across the filters by state-of-the-art instrumentation provides partial energy savings by avoiding putting dirty filters online, and hence avoiding shorter run times (Daffer, 1984). The rate of head loss buildup is a function of the size of the media. The smaller the media, the higher the initial head loss and rate of head loss buildup. This is even more true with a media such as GAC, where the size of the media selected will directly influence the cost of energy and availability of head. This also has an effect on the additional energy required to backwash such a filter as compared to a sand filter. Additionally, GAC filters have been found to be more energy-intensive than sand filtration because the media has to be regenerated periodically. The maximum total energy required for on-site regeneration of activated carbon is estimated to be 8300 Btu/lb (2.43 kWh/lb) (includes furnace fuel, steam, fuel for an afterburner, and so on). Alternative technologies for the removal of taste and odor compounds as well as organics from water should be researched and considered whenever possible (Daffer, 1984).

With a constant-rate filtration system, the clearwell is generally constructed underground to allow gravity flow; however, for declining rate filters, it is possible to raise the clearwell to an elevation that is just sufficient to keep the media submerged at all times. This can provide energy savings by reducing the total static head that the finished water high-service pump has to pump against (Daffer, 1984).

The operator can also achieve additional energy savings by scheduling the backwash cycle for periods of lower energy cost. Because pumps represent the primary energy users for traditional filters, utilities need to consider the potential savings by selecting and sizing the proper pumps and using high-energy efficiency motors. Alternatively, use of elevated storage tanks for filter backwash water can minimize the energy required for backwashes by allowing the tank to be filled during low-demand/off-peak energy rate times.

6.2   Membrane Filtration

6.2.1   Low-Pressure Membrane Filtration (Microfiltration/Ultrafiltration)

Low-pressure membrane filtration is becoming increasingly popular with water treatment facilities due to the high-quality water it can produce, the reliability of the process, and its smaller footprint. Microfiltration and ultrafiltration are two types of low-pressure membrane processes that are characterized by their ability to reject suspended and colloidal particles via a sieving mechanism. Low-pressure membrane systems are typically configured as submerged systems such as General Electric Water & Process Technologies/ZENON Environmental Inc.'s (Trevose, Pennsylvania) ZeeWeed ultrafiltration membranes or encased such as Hydranautics' (Oceanside, California) HYDRAcap®.

Submerged membrane systems are those where the membrane fibers are immersed in an open tank and fed by gravity. Groups of membrane fibers are bundled together in racks or modules. During normal operation, coagulated water enters the process tank and completely submerges the membrane modules. The raw water is drawn through the membrane fibers (outside-in) to the inside of the fiber by applying a small vacuum 55 to 83 kPa (8 to 12 psi). Filtrate exits through the top of the module to the permeate manifold. Generally, multiple modules are connected to one permeate pump through a common manifold. The outside-in scheme has the advantage of a larger membrane surface area, which allows for a slightly higher flow than the inside-out (where the water flows through the inside of the fiber to the outside) model while still maintaining the same flux rate and solids concentration.

Conversely, encased membrane systems use the same hollow fibers as the submerged systems except that they are packed into a cylindrical casing, usually with a diameter of 200 mm (8 in) and a length of 1000 mm (40 in). These fiber packages, or elements, are then arranged end-to-end in a pipe or vessel. In ultrafiltration applications, the vessels are commonly oriented vertically; each vessel houses one element. For standard pressure-driven systems, the pressure drop across the system is generally on the order of 210 kPa (30 psi).

There are two types of filtration modes: (1) dead end, where the flow path is perpendicular to the membrane surface and all the water is filtered, and (2) cross flow, where the flow path is parallel to the membrane surface and only a portion of the flow is filtered with the remaining flow recycling back. Because water is pumped through the membrane vessel, the cross-flow mode uses more energy than the dead end mode. However, in the cross-flow mode, solids are continuously flushed from the system, resulting in less frequent back pulses and backwashes and possibly longer membrane life. In dead-end operational mode, solids build up in the system, thus more frequent backwashes and back pulses will be required.

The City of Lake Forest, Michigan's, water facility has a design capacity of 53 ML/d (14 mgd) at 20 °C. The temperature of the raw water varies from 0.1 to 26 °C (32 to 79 °F). The average daily water demand is around 15 ML/d (4 mgd). The plant's design features seven membrane units. Two operation philosophies were possible: (1) operating the minimum number of units in cross-flow mode and (2) operating all seven units in dead-end mode. Selecting the second alternative provided energy savings of 333 J/L (350 kWh/mil gal). Like conventional media filtration, low-pressure membrane filtration also requires periodic backwash (both air and/or water) based on a set transmembrane pressure or time interval. Backwashing or back-pulse type (some manufacturers use chemically enhanced backwashes) and frequency are manufacturer-specific.

A 2001 AwwaRF study that compared UV disinfection for *Cryptosporidium* inactivation with low-pressure membrane filtration and ozone reported membrane filtration as having energy consumption ranging from 143 to 480 J/L (150 to 500 kWh/mil gal), corresponding to encased filtration in dead-end mode and submerged systems, respectively (American Water Works Association Research Foundation, 2001).

The following is a hypothetical example to illustrate the various components of a submerged low-pressure membrane system and associated energy use for a 10-mgd surface water facility. The following assumptions are made about the facility: 40-ML/d (10-mgd) net permeate production; treating clarified water (turbidity less than 2 NTU, 95% of the time); water temperature of 15 °C; 95% recovery; four treatment trains with two submerged cassettes per train; permeate is discharged to a below-ground clearwell (minimal pumping head); and treatment trains operating at average trans-membrane pressure values. Table 6.5 provides a list of major equipment and connected motor loads as well as daily power consumption under average conditions.

The power consumption for a low-pressure membrane filtration system is a function of plant flow variations, raw water quality, pretreatment processes, influent water quality, water temperature, plant configuration, desired treated water quality goals, and will vary from system to system.

TABLE 6.5   Typical low-pressure system average power consumption (10-mgd surface water plant) (courtesy of GE Water & Process Technologies/ZENON Environmental Inc.)

Major Equipment	Qty.	Motor size (hp)	Starter type	Connected load (hp)	Power consumption kWh/day
Permeate pumps	4	30	VFD	120	1261
Backwash pumps	2	30	VFD	60	13
Air blowers	2	15	VFD	30	18
Clean-in-place pump	2	5	VFD	10	11
Cleaning tank heater					116
Heat recirculation pumps	2	3	FVNR	6	32
Air compressor	2	7.5	FVNR	15	13
Total average power consumption for the system					1464

VFD = variable frequency drive
FVNR = full voltage non-reversing

6.2.2 Low-Pressure Reverse Osmosis and Nanofiltration (Brackish Water Desalination)

Low-pressure reverse osmosis (LPRO) and nanofiltration are semi-permeable membranes that reject dissolved solids in addition to particulate matter. They are used to produce potable water from a source water that exceeds the drinking water standards for total dissolved solids (TDS) and certain dissolved solids such as hardness ions or chlorides and sulfates (LPRO/nanofiltration is used for low-to-high brackish water sources where TDS ranges from 400 to 3000 mg/L). Pressure is applied on the feed side of the membrane to force water through the membrane while retaining dissolved solids. The applied pressure is a function of the osmotic pressure of the water that is directly proportional to the concentration of dissolved solids in the feed water. Hence, 90% of the energy consumption in a LPRO/nanofiltration process results from the pressure applied on the membranes to produce a certain quantity of water. The main difference between LPRO and nanofiltration membranes is the lower rejection capabilities of monovalent ions by the nanofiltration membrane. Typical operating pressures of LPRO/nanofiltration membranes range from 480 to 1400 kPa (70 psi to 200 psi). As a general rule, use 7 kPa (1 psi) of pressure for every 100 mg/L of TDS differential between the feed/concentrate and permeate side. This translates into energy requirements of 665 to 1900 J/L (700 to 2000 kWh/mil gal) for a system recovery of 85%.

Depending on raw water quality, membrane rejection capabilities, and finished water quality goals, membrane processes such as LPRO can allow for a percentage of the raw water to be blended with the process product water to minimize the size of the membrane process. Because little or no treatment is required for the raw water blend stream, this water represents 100% recovery and is a low cost means to increase overall capacity or, equivalently, decrease the net required installed capacity for the membrane process. Blending is possible if the membrane process removes sufficient TDS constituents to a level that allows a percentage of raw water to be combined with the process product water without exceeding finished water quality goals.

7.0   DISINFECTION

7.1   Chlorine Gas

Liquid chlorine is delivered in ton cylinders. For withdrawal rates beyond that which the transfer of heat from the ambient air can support, liquid chlorine is evaporated using an electrically heated evaporator. The evaporator automatically vaporizes and superheats liquid chlorine at a rate controlled by system demand. The water heater is typically an electric-immersion type (18-kW maximum for a 4500-kg [10 000-lb] unit). Chlorinators are vacuum-operated devices used to control chlorine gas feed rates. The chlorine gas control system operates under a vacuum to prevent gas leakage. This method of chlorination is insignificant from an energy consumption standpoint. However, high-pressure plant water is required to produce the required vacuum to dissolve gaseous chlorine in water and convey it to the injection point. The quantity of motive water and pressure are a function of the size of the eductor, the backpressure on the solution line, and the chlorine feed rate.

7.2 Bulk Sodium Hypochlorite

Commercially purchased bulk sodium hypochlorite is delivered at 10 to 12% concentration to bulk storage tanks. Metering pumps feed the sodium hypochlorite to injection points within the plant. Because the metering pumps use fractional horsepower, this method of chlorination is insignificant from an energy consumption standpoint.

7.3 On-Site Hypochlorite Generation

With on-site hypochlorite generation equipment, sodium hypochlorite is produced and stored at the plant site using an on-site generator. Concentrated brine solution made from food-grade sodium chloride (NaCl) and softened water passes through the electrolytic cells producing 0.8% sodium hypochlorite and hydrogen gas. The hypochlorite is metered into the process flows and the hydrogen gas is vented to the atmosphere. The generation system requires 1.58 kg (3.5 lb) of solar salt, 56.7 L (15 gal) of water, and 9 MJ (2.5 kWh) of electricity to produce 0.5 kg (1 lb) of equivalent chlorine as a 0.8% sodium hypochlorite solution. Therefore, a chlorine dose in the range of 2 to 8 mg/L requires 7.7 to 31 kg/day (17 to 68 lb/day) per milligram of chlorine demand. The energy consumption for producing the required sodium hypochlorite is in the range of 40.4 to 160 J/L (42.5 to 170 kWh/mil gal). Because the hypochlorite can be stored in day tanks, energy savings can be achieved by producing the required hypochlorite during off-peak energy rate hours.

7.4 Ultraviolet Systems

Ultraviolet disinfection is primarily used in the water industry for disinfection. Ultraviolet light is a result of excitation of mercury electrons. An applied voltage differential across a mercury-argon gas mixture results in discharge of photons or light that is in the germicidal range (wavelength of 200 to 300 nm). Ultraviolet light damages DNA/RNA, making it noninfectious. Ultraviolet light intensity is described in terms of milliwatts per square centimeter and UV dose in terms of milliwatt seconds per square centimeter. Ultraviolet dose delivery is a function of lamp intensity and velocity distribution through the pipe. Optimization of dose monitoring and control has a significant impact on electrical power consumption and lamp replacement. Effective filtration reduces the quantity and size of particles present, which leads to improved UV system performance. Filtration has a direct effect on the ultraviolet transmittance (UVT) of water. Ultraviolet dose can be expressed as follows:

$$\text{Ultraviolet delivered dose} = \text{EOLL} \times \text{FF} \times \text{function}$$
$$\text{(Head loss, UVT, Power setting, and Flow)} \quad (6.3)$$

Where:

EOLL = end of lamp life;
FF = sleeve fouling factor; and
UVT = ultraviolet transmittance.

Any change in the aforementioned parameters affects the delivered UV dose and, as a result, energy consumption. Figure 6.10 shows the effect of UVT on delivered dose and energy consumption (data is based on an equation developed for a low-pressure high-output [LPHO] system from validation testing done by Carollo Engineers [Phoenix, Arizona]).

Ballasts are transformers that control the power to ultraviolet lamps. Ballasts should operate at temperatures below 60 °C to prevent premature failure. Two types of transformers are commonly used with UV lamps: electronic and electromagnetic. Electronic ballasts operate at a much higher frequency than electromagnetic ballasts, resulting in lower lamp-operating temperatures, less energy use, less heat production, and longer ballast life.

Ultraviolet lamps used in full-scale facilities are manufactured as low-pressure, low-pressure high-output, and medium-pressure lamps. Table 6.6 provides inherent characteristics of the three different mercury vapor lamps, which show that low-pressure lamps are the most energy-efficient lamps. However, depending on the size of the water facility, a life cycle analysis should be done prior to selecting the type of lamp.

A 2001 AwwaRF study that compared UV disinfection for *Cryptosporidium* inactivation (a UV dose of 40 MJ/cm^2) with low-pressure membrane filtration and ozone reported UV disinfection as having the lowest energy consumption of all, ranging from 48 to 140 J/L (50 to 150 kWh/mil gal), corresponding to LPHO and

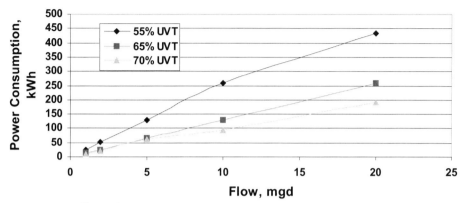

FIGURE 6.10   Effect of UV transmittance on power consumption.

Table 6.6   Ultraviolet lamp characteristics (U.S. EPA, 2006)

Parameter	Low-pressure (LP)	Low-pressure, high output (LPHO)	Medium pressure (MP)
Germicidal UV light	Monochromatic at 254 nm	Monochromatic at 254 nm	Polychromatic at 200–300 nm
Emission	Continuous-wave	Continuous-wave	Continuous-wave
Mercury vapor pressure (Pa)	Approximately 0.93 (1.35×10^{-4} psi)	0.18–1.6 (2.6×10^{-5}–2.3×10^{-4} psi)	40 000–4,000,000 (5.80–580 psi)
Operating temperature	Approximately 40°C	60–100°C	600–900°C
Electrical input (W/cm)	0.5	1.5–10	50–250
Germicidal UV output (W/cm)	0.2	0.5–3.5	5–30
Electrical to germicidal UV conversion efficiency (%)	35–38	30–35	10–20
Arc length	10–150 cm	10–150 cm	5–120 cm
Relative number of lamps needed for given dose	High	Intermediate	Low
Lifetime (hr)	8 000–10 000	8 000–12 000	4 000–8 000

medium-pressure UV systems (American Water Works Association Research Foundation, 2001). The study also revealed that LPHO lamps consumed the least energy per lamp compared to low-pressure and medium-pressure lamps. Additionally, head loss, which is a function of UV reactor design, is at a minimum when the flow is parallel to the UV lamps than when the lamps are oriented perpendicular to the flow path (American Water Works Association Research Foundation, 2001).

Another study conducted by AwwaRF in collaboration with NYSERDA and the cities of Phoenix, Arizona, and Tacoma, Washington, suggests use of advanced battery systems that could potentially save electricity for energy-intensive processes such as UV disinfection (American Water Works Association Research Foundation and New York State Energy Research and Development Authority, 2007). These advanced battery systems, such as a sodium-sulfur battery that has been developed in Japan and is now commercially available, have been shown to provide energy savings in that they can be charged during off-peak hours (mostly at night) and later run UV systems during the day when electricity charges are much higher (U.S. EPA, 2006). Utilities that can purchase electricity at a variable-rate structure can benefit from this approach. However, a drawback of this approach at present is the high capital cost associated with the batteries. There is general consensus that with widespread use of these batteries, which can also be used as uninterruptable power supply units for sensitive equipment/processes, prices would become more affordable over time.

Energy savings could also be achieved by "pacing" the dose with flow by turning lamps on/off with flow variations. However, it has been found that this may pose a complex operational scheme. Care should also be taken to avoid addition of any chemicals such as corrosion inhibitors or other water conditioning chemicals upstream of the UV system. Such chemicals could coat the UV apparatus and also reduce UV transmittance.

8.0 HIGH-SERVICE PUMPING

Treated water is typically stored on-site in a ground storage tank and is pumped off-site to either elevated storage tanks or ground storage tanks within the distribution system. As shown in Figure 6.2, finished water pumping accounts for more than half of the energy consumption in a water plant. The location of a water plant and topography govern the energy required to pump the water. Many utilities have failing infrastructure with aging and corroding pipes that require more energy to transport

water. Water lost during transmission through leakage is energy wasted. It is estimated that 18 to 36 billion MJ (5 to 10 billion kWh) of power generated in the United
States is expended each year on water that is unaccounted for, that is, either leaked
away or not paid for by the customer (American Water Works Association, 2003).
Aging infrastructure also adversely impacts water quality. U.S Environmental Protection Agency's Stage 2 Disinfectants and disinfection byproducts rule, as well as
increasing customer expectations regarding taste and odor of treated water, requires
utilities to reevaluate strategies on energy-saving techniques such as off-peak
pumping and providing large storage volumes, for extended periods of time, within
the distribution system away from the treatment plant.

Treated water is typically stored in ground storage tanks and then transferred to
elevated storage tanks via pumping when the water level in the elevated storage tank
falls below a preset point. By lowering the preset points during periods of low water
consumption so that the pump is activated fewer times but runs longer is one of the
methods that can be implemented to save energy (Clark, 1987).

By tracking individual energy use for a set of pumps operating for the same
application (adding submetering on each pump), operators can optimize the
pumping operation. In Longview, Texas, where the principal function of the
pumping system was to pump to elevated storage, kilowatt-hours/million gallons
were tracked over a 4-year period and ranged from nearly 2200 J/L (2300 kWh/mil.
gal) during low consumption periods to less than 1140 J/L (1200 kWh/mil gal)
during months of high consumption. This comparison indicated that the system was
inefficient during periods of low water use, thus instigating an investigation to how
system operation could be improved. By making more effective use of storage, operations were improved so that year-round performance averaged about 1050 J/L (1100
kWh/mil gal) (Clark, 1987).

Several energy conservation measures are explained in Section 1.0 of this chapter.
Additionally, energy conservation measures relevant to high-service pumping are
described in Chapters 3, 4, and 5 of this manual.

9.0   WATER PLANT RESIDUALS MANAGEMENT

Residuals in a water facility depend on the processes used for treatment. Typically, surface water facilities using filtration would have two significant sources for residuals
from the coagulation/flocculation/sedimentation process: the spent filter backwash
water and the filter-to-waste water. The coagulation/flocculation/sedimentation

process residuals are typically scrapped from the bottom of the sedimentation tank and pumped to either a gravity thickener or a holding tank, and later thickened and dewatered for disposal. The spent filter backwash water is held in an equalization tank and the overflow is pumped back to the head of the plant for reprocessing. Typically, spent filter backwash volumes vary from 2 to 6% of the plant capacity. However, it depends on the size of the plant, number of filters, and frequency of filter backwashes per day. It should be noted that upstream treatment processes affect the loading of solids on the filters and, thus, the frequency of filter backwashes.

9.1   Gravity Thickeners

Gravity thickeners are essentially clarifiers with mechanically driven scrappers to thicken sludge. The torque applied is higher than a typical secondary clarifier in an activated sludge plant due to higher sludges that need to be handled by the rake assembly. The mechanical power required to drive these units is small (normally <1.5 kW [2 hp]) and, therefore, an insignificant consumer of electricity. However, the entire thickening process would include the pumps pumping from the clarification process to the gravity thickeners, the thickener overflow pumps pumping the overflow back to the head of the plant, and the thickened sludge pumps to a day tank or holding tank. Energy consumption is in the range of 6.7 to 87.5 J/L (7 to 92 kWh/mil. gal) (flow is plant design flow) for the gravity thickening process, including all associated pumping. Data for the energy numbers were obtained from a project initiated by the American Water Works Association (AWWA) (Denver, Colorado) to come up with cost estimates for implementing residuals management at existing coagulation plants to assist U.S. EPA on section 304(b) of the Clean Water Act (Roth et al., 2008).

9.2   Belt Filter Press

A belt filter press uses the dewatering physics of gradually applied mechanical pressure; that is, the more liquid removed from the sludge, the more stable that sludge becomes and the higher the pressure it can withstand. Ultimately, a drier cake results. This method of gradual dewatering is accomplished by using drainage, low-pressure, and high-pressure dewatering zones. For aid in dewatering, a polymer is mixed with the sludge upstream of the belt filter press. Typical solids application rates for coagulation residual (alum or ferric sludge) are in the range of 140 to 204 kg/m·h (300 to 450 lb/hr/m). Sludge is pumped from either a holding tank or a gravity thickener to the belt filter press. The belt filter press is estimated to achieve a dry cake of 12 to

18% solids. The energy requirement is 49.6 kJ/kg (12.5 kWh/ton) of dry solids. The belt filter press needs to be cleaned continuously with wash water. The wash water requirement averages 150 to 420 L/min (40 to 110 gpm) at 550 to 900 (80 to 130 psi), which consumes 3.73 to 10.44 kW (5 to 14 hp) of energy consumption.

9.3 Centrifuges

The centrifuge uses the principle of accelerated settling to separate solids from a liquid. Feed is introduced through a stationary feed tube into a rotating bowl, where G-force causes the solids to "sediment" against the bowl wall. The action of a screw conveyor mounted concentrically within the bowl transports the solids up an incline and out of the machine. The effluent flows axially toward the front of the machine and is discharged over adjustable weirs. For flocculation, a polymer is mixed upstream of the centrifuge, preferably on the discharge side of the feed pump or at the feed tube. The centrifuge is estimated to achieve a dry cake of 18 to 25% solids. The energy requirement is in the range of 9.5 to 71.3 J/L (10 to 75 kWh/mil. gal) (flow is treatment plant design flow). The energy requirement includes feed pumps (pumping thickened sludge) to the centrifuge and centrate recycle pumps back to the plant. Data for the energy numbers were obtained from the same previously mentioned AWWA-initiated project to come up with cost estimates for implementing residuals management at existing coagulation plants to assist U.S. EPA on section 304(b) of the Clean Water Act (Roth et al., 2008).

9.4 Membrane Concentrate Disposal

Concentrate generated from membrane processes, especially reverse osmosis/ nanofiltration membranes, is typically disposed of as a surface water discharge, treated at the wastewater facility, disposed of by deep-well injection, land-applied (via spray irrigation), or evaporated in solar ponds. A survey of existing full-scale facilities conducted by the U.S. Department of the Interior, Bureau of Reclamation (2001), showed that about 85% of desalination facilities dispose their concentrate to some sort of surface water. Therefore, energy requirements for surface water discharge would represent pumping costs associated with pumping the concentrate from the facility to the permitted surface water source. Few facilities use energy-intensive zero-liquid-discharge thermal processes such as evaporators, concentrators, spray dryers, and crystallizers. It takes about 1000 kJ (1000 Btu) to evaporate 0.5 kg (1 lb) of water. Electric power consumption for brine concentrators can range from

57 to 95 kJ/L (60 to 100 kWh/1000 gal) of feed water; vapor compression crystallizers can range from 190 to 240 J/L (200 to 250 kWh/1000 gal) of feed water; and, for spray dryers using oil and natural gas, consumption is in the range of 12 kJ/L · s (0.7 Btu/gpm) (U.S. Department of the Interior, Bureau of Reclamation, 2001).

10.0 REFERENCES

American Water Works Association (2003) Water Loss Control Committee Report: Applying Worldwide BMPs in Water Loss Control. *J. Am. Water Works Assoc.*, **95** (8), 65.

American Water Works Association Research Foundation (2008) *Risk and Benefits of Energy Management for Drinking Water Utilities;* American Water Works Association Research Foundation: Denver, Colorado.

American Water Works Association Research Foundation (2003) *Improvement of Ozonation Process through Use of Static Mixers.* American Water Works Association Research Foundation: Denver, Colorado.

American Water Works Association Research Foundation (2001) *Practical Aspects of UV Disinfection;* American Water Works Association Research Foundation: Denver, Colorado.

American Water Works Association Research Foundation; New York State Energy Research and Development Authority (2007) *Optimization of UV Disinfection;* American Water Works Association Research Foundation: Denver, Colorado.

California Energy Commission (2005) *California's Water-Energy Relationship;* CEC-700-2005-011-SF; California Energy Commission: Sacramento, California.

Clark, T. (1987) Reducing Power Costs for Pumping Water. *Opflow,* **13** (10).

Daffer, A. R. (1984) Conserving Energy in Water Systems. *J. Am. Water Works Assoc.*, December, 34–37.

Energy Center of Wisconsin (2003) *Energy Use at Wisconsin's Drinking Water Facilities;* ECW Report No. 222-1; Energy Center of Wisconsin: Madison, Wisconsin.

Electric Power Research Institute (2002) *Water & Sustainability: U.S. Electricity Consumption for Water Supply & Treatment—The Next Half Century;* Product ID #1006787; Electric Power Research Institute: Palo Alto, California.

Electric Power Research Institute (1999) *A Total Energy & Water Quality Management System*; Product ID #TR-113528; Electric Power Research Institute: Palo Alto, California.

Electric Power Research Institute (1996) *Water and Wastewater Industries: Characteristics and Energy Management Opportunities*; Product ID #CR-106491; Electric Power Research Institute: Palo Alto, California.

Electric Power Research Institute (1994) *Energy Audit Manual for Water/Wastewater Facilities*; CEC Report CR-104300; Electric Power Research Institute: Palo Alto, California.

Jacobs, J.; Kerestes, T. A.; Riddle, W. F. (2003) *Best Practice for Energy Management*; American Water Works Association Research Foundation and American Water Works Association: Denver, Colorado.

Kavanaugh, M. C., and Trussell, R. R. (1980) "Design of aeration towers to strip volatile contaminants from drinking water" *Journal American Waterworks Association*, Research and Technology, Vol. 72, No. 12., pp. 684 - 692.

Lalezary, S.; Pirbazari, M.; McGuire, M. J.; Krasner, S. W. (1984) Air Stripping of Taste and Odor Compounds from Water. *J. Am. Water Works Assoc.*, March, 83–87.

Layne, J.; Eckenberg, W. G. (1983) Denver Foothills Project: Energy Efficiency in Action. *J. Am. Water Works Assoc.*, October, 487–491.

Rackness, K.L. (2005) *Ozone in Drinking Water Treatment - Process Design, Operation, and Optimization.* AWWA TD461.R35, Denver, CO

Rackness, K. L.; Russell, R.; Gifford, G.; Zegers, R.; Hunter, G. (2000) Ozone Control at Las Vegas to Obtain On-peak, Off-peak Energy Savings. Proceedings of the Pan American Group Conference, October 2000. International Ozone Association, Orlando, Florida.

Rackness, K. L.; DeMers, L. D. (1998) *Ozone Facility Optimization Research Results and Case Studies*; AwwaRF and EPRI, California Energy Commission (CEC), St. Louis, Missouri.

Roth, D. K.; Cornwell, D. A.; Russell, J.; Gross, M.; Malmrose, P.; Wancho, L. (2008) Implementing Residuals Management: Cost Implications for Coagulation and Softening Plants. *J. Am. Water Works Assoc.*, March, 81–93.

Trivedi, R.; Van Cott, W.; Stevenson, R.; Cogswell, T. M.; Toledo, T. P. (1995) Streamlines Treatment Operations. *J. Am. Water Works Assoc.*, August, 34–42.

U.S. Environmental Protection Agency (2006) *Ultraviolet Disinfection Guidance Manual for the Final Long Term 2 Enhanced Surface Water Treatment Rule;* EPA-815-R-06-007; U.S. Environmental Protection Agency: Washington, D.C.

U.S. Environmental Protection Agency (2008) *Ensuring a Sustainable Future: An Energy Management Guidebook for Wastewater and Water Utilities;* U.S. Environmental Protection Agency: Washington, D.C.

U.S. Department of the Interior, Bureau of Reclamation (2001) *Membrane Concentrate Disposal: Practice and Regulations;* U.S. Department of the Interior, Bureau of Reclamation: Washington, D.C.

Chapter 7

Energy Use in Wastewater Treatment Processes

Energy use may vary significantly for a given-size wastewater treatment plant (WWTP), depending on its location, strength of wastewater, level of treatment, in-plant recovery, type of treatment process selected, and mode of operation. Therefore, it is not practical to provide a general energy usage level for WWTPs. Generally, conventional aquatic lagoon systems and land treatment systems are among the least energy intensive. Advanced treatment processes often require greater amounts of primary energy and also require greater amounts of chemicals, with associated secondary energy use. Factors such as wastewater strength, hydraulic conditions, in-plant energy-recovery design, and operation mode greatly affect electrical power and fossil fuel energy uses. Figure 7.1 shows typical profiles for annual energy use for different types of secondary treatment WWTPs with various treatment processes at 0.44-m^3/s (10-mgd) average flow.

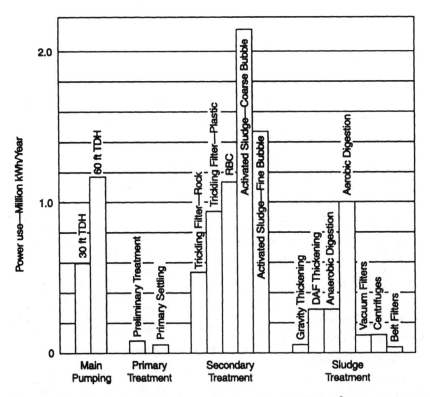

FIGURE 7.1   Typical energy use profile for 10-mgd (0.4-m^3/s) secondary treatment processes.

The energy uses of WWTPs are generally considered to be those directly associated with on-site electrical power and fuel requirements. These are typically referred to as *primary energy uses.* Secondary energy uses are associated with materials manufacture and transport and manufacture of consumable chemicals that are used in the treatment process. A discussion of energy use for specific unit processes follows.

1.0   PRELIMINARY AND PRIMARY TREATMENT

Preliminary treatment is a low-energy-intensity area of the wastewater treatment process; however, if not properly designed and operated, it could waste important hydraulic head and potentially increase the amount of pumping required in the treatment process. This section provides information related to energy conservation. The reader is also referred to the U.S. Environmental Protection Agency's (U.S. EPA's) Wastewater Technology Fact Sheet: Screening and Grit Removal (http://www.epa.gov/owmitnet/mtb/screening_grit.pdf) for an overview of screening and grit removal processes.

1.1   Screening

Screens are typically the first treatment devices in WWTPs. Screening units are cleaned either manually or by mechanically driven devices. Manually cleaned screens are usually limited to bar racks before lift stations at small plants. Mechanically cleaned bar screens are used for all other applications. In general, it is not necessary to operate the cleaning rakes continuously, and most applications include adjustable timing devices or differential water level devices to activate the screen-cleaning mechanism.

The past 10 years have seen improvements in the area of wastewater screening. With the advent of membrane filtration and improvements in maintenance associated with improved screening, equipment is now available on the market that includes a selection of opening sizes as small as 1 to 6 mm. Other improvements include the development of screening conveyors that wash and compact the screened material. In some instances, the screened material is packaged into plastic bags for landfill disposal.

Influent-screening energy consumption is primarily due to the screen rake or band screen motor, whose operation is driven by head losses across the screen. Screening represents a minor portion of total WWTP energy use, with units handling $0.7 \text{ m}^3/\text{s}$ (15 mgd) or less typically being driven by 0.55-kW (0.75-hp) motors.

Head loss through the screens requires additional pumping. The hydraulic profile should be analyzed and, if appropriate, the timing cycle or differential water level used to initiate screen cleaning should be monitored and adjusted to prevent excessive losses.

Energy savings opportunities exist via the reduction in wash water flows for rinsing of screenings. Solenoid valves should be installed and maintained to shut the wash water flow off when the screen is not in the process of being cleaned. Effective screening can substantially reduce the energy required in other parts of the plant.

1.2 Influent Wastewater Pumping

Depending on the WWTP site elevation and influent sewer elevation, the influent wastewater pumping energy requirement alone can represent 15 to 70% of the total WWTP use of electrical energy. Moreover, if the energy required to operate all of the pumps in the collection system is considered, total pumping energy requirements may represent as much as 90% of the total energy used (see Chapter 4).

For a typical direct-drive influent pump with a total dynamic head (TDH) of 9 m (30 ft), a pump efficiency of 74%, and a motor efficiency of 88%, the pump wire-to-water power requirement would be 1.6 W/m^3 · d (8.1 hp/mgd) or 6.0 kW/mgd (mgd × [4.383 × 10^{-2}] = m^3/s) (see Chapter 2). Estimating power requirements by multiplying the influent flow received by the pump system energy ratio is much more accurate than using the motor horsepower and estimated run times (see Chapter 4 for a more complete description of energy requirements for pumping).

As an example, assume that a station with two 75-kW (100-hp), 0.5-m^3/s (12-mgd) pumps and a design TDH of 9 m (30 ft) processes 0.8 m^3/s (18 mgd) over a 24-hour period. What is the estimated average power draw and energy requirement?

Pump capacity (both pumps) = 24 mgd

Flow = 18 mgd

Operating hours = (18 mgd/24 mgd) × 24 hours = 18 hrs/day

Power (both pumps) = 75 kW + 75 kW = 150 kW

Energy required = 150 kW × 18 hr/day = 2700 kWh/day

1.2.1 Opportunities for Energy Reduction

Energy use at an influent pumping station is determined by how much of the time the pumps operate at their best efficiency point. While computation of energy requirements is a standard calculation, designs can be developed that optimize pumping power.

Use of higher efficiency pumps and motors can be cost-effective, even if initial costs are greater than their less-efficient counterparts. For the aforementioned example, selection of a high-efficiency motor (94% versus 88%) would reduce the wire-to-water power requirement from 6 kW/mgd to 5.6 kW/mgd.

The use of variable-frequency drives (VFDs) has been shown to be a more efficient use of power. The advantage of VFDs is that the pump can operate at its optimum efficiency for the entire flow range. Pumps should be selected that stay within their optimum efficiency range over the system curve with reduced speeds.

Standard design suggests that pumps should be sized to handle peak hour flows to the treatment plant with the greatest head conditions. The static head condition in design is taken as the differential between the lowest wet well level and the highest discharge elevation.

As illustrated in Figure 7.2, if the normal high wet well level is used as the design condition for peak hour flow, it represents a condition that can be accommodated for this time period without effect on the station; this reduces the design

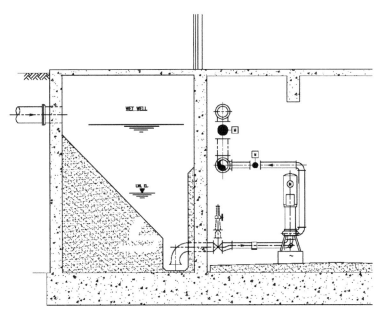

FIGURE 7.2 Wet well operation (courtesy of Metcalf & Eddy, 2004).

condition significantly. In the aforementioned example, if design static lift is reduced from 5 m to 4 m, peak hour power draw would be reduced by more than 10%.

1.3   Grit Removal

Although grit removal is not a process that consumes a great deal of energy, there is some energy use involved and thus some room for conservation. Effective grit removal can substantially reduce the energy required in other parts of the plant. If not properly removed, it can accumulate in anaerobic digesters, reduce effective digester volume, and reduce the production of digester gas, a renewable resource. There are also a number of different styles of grit removal chambers, each relying on varying degrees of energy input to foster removal of grit and retain organic material. The type of device/system selected during design will largely determine the energy intensity of the process.

Many newer grit basins use the vortex design to collect grit, with pumps used to transport the grit to a hydroclone separator where the grit is concentrated and discharged to a grit washer prior to discharge to a dumpster for disposal in a landfill. The major source of energy is in the pumps used to transport the grit, although small motors may be used on center and conveyor drives. Energy intensity can be controlled by operating grit pumps on cycles that prevent buildup of grit in the tanks without operating continuously. Grit washer and conveyor drives can be connected through interlocks to operate only when the pumps are operated and for some reasonable period of time after the pumps are shut off. Premium efficiency motors should be carefully considered when the motors on the grit pumps need replacement.

Aerated grit chambers use blowers to introduce air that creates turbulence to suspend lighter organic material and allows heavier grit particles to settle. The key to optimal operation of aerated grit chambers is to match the air produced with the air required for optimal operation. Too little air results in putrescible material settling at the bottom of the tank, producing odors and making storage and disposal of grit more difficult. Too much air will result in too little grit being removed, placing a greater burden on downstream unit processes and sludge disposal operations.

Matching blower output with necessary air requirements is a trial-and-error process. There is no theoretical way to exactly determine the blower output required. Optimal output can only be determined by repeated tests and will vary with the time of day, season, and during wet weather events. In installations with variable-speed drivers for blowers, blower output can be matched with air requirements and programmed into the plant's process control system. If changing the air flow requires changing the driver, an economic analysis should be done to determine whether this change is justifiable.

Early grit basins were designed to maintain a minimum velocity, allowing grit to settle and organics to be retained in the process stream. In velocity-type grit basins, little energy is used and, once the basin is constructed, not much additional energy can be saved. The choice of grit removal process can also impact the hydraulic profile of the plant and thus the overall energy requirements of the facility.

2.0   PRIMARY TREATMENT

The purpose of primary treatment is to remove settleable solids and floating material from wastewater. An easily overlooked impact of primary treatment is the ensuing cost of secondary treatment, which is affected by removals in primary treatment. Poor primary removals negatively affect the energy footprint of the overall plant by: (1) requiring greater use of aeration energy in the secondary process; (2) requiring more energy in solids handling and disposal as a result of the production of a greater secondary-sludge-to-primary-sludge ratio; and (3) reducing digester gas recovery in plants with anaerobic digesters.

When examining energy uses in primary treatment, it is useful to consider conventional primary treatment separately from the lesser used chemically enhanced primary treatment. Chemically enhanced primary treatment involves essentially all of the direct energy uses in conventional primary treatment in addition to direct energy uses associated with chemical feed, mixing, and pumping. Primary sludge handling will be discussed as a standalone topic because it is common to both conventional and chemically enhanced primary treatment.

2.1   Conventional

Primary settling tank energy uses may include drives for sludge and skimmings' collector mechanisms, skimmings' pumps and grinders, channel aeration blowers, and gallery exhaust and supply fans. Of these energy uses, blowers and fans often afford the greatest opportunity for energy savings. If variable-speed/frequency drives on channel air blowers are present an opportunity may exist for minimizing energy use while maintaining solids suspension or other process functions of the channel air. Timers or remote switches on the gallery exhaust and supply fans may reduce energy costs of fans that are constantly operating, although safety codes for the region must be adhered to and take priority over energy savings. At a minimum, the National Fire

Protection Association's (Quincy, Massachusetts) (2008) *Standard for Fire Protection in Wastewater Treatment and Collection Facilities* should be referenced.

2.2   Chemically Enhanced

Chemicals can be used to enhance primary treatment. This process enhancement results in increased biochemical oxygen demand (BOD)/chemical oxygen demand (COD) and total suspended solids removals and reduces the organic loading to the secondary processes. This reduced loading results in a reduced energy requirement, which needs to be accounted for in the development of operation and maintenance costs or energy balances. Direct energy uses in this type of process may include rapid (flash) mixing, polymer mixing, polymer transfer pumps, polymer addition pumps, chemical mixing, chemical transfer pumps, and chemical addition pumps. The greatest amount of energy is most likely used in the rapid (flash) mixing process. A discussion of rapid (flash) mixing theory is presented in Section 4.0 of Chapter 6. Although this discussion relates specifically to water treatment, the general aspects of mixing energy also apply to wastewater. If energy use in this process is substantial, it may be prudent to conduct an examination to establish the most cost-effective balance of energy use, chemical consumption, and process removals. From the perspective of a total energy footprint, one could also consider energy used in the manufacture and delivery of the chemicals themselves, although financially this would be accounted for in their purchase price.

2.3   Primary Sludge Pumping

The primary sludge pumping operation is generally the largest energy-consuming component of a conventional primary treatment process. Energy uses may include primary sludge pumps, primary sludge grinders, and primary sludge valves. Because these energy uses are process driven and established by design constraints such as flow and head requirements on the pumps, they do not readily afford opportunities for energy savings. However, some plants have had success in reducing overall process costs by a selective approach to intermittent/modulated primary sludge pumping, which facilitates pumping a reduced volume of a more highly concentrated sludge to digesters. Installation of VSDs or variable-speed drives in conjunction with density meters can facilitate this pumping approach. The benefits of intermittent/modulated pumping are only realized if pumping stops when a drawdown cone appears in the primary clarifier sludge blanket. If a plant's process train and clarifier configuration can accommodate this approach while still removing sufficient sludge, the advantages may include less energy use for sludge pumping and lower volumetric loading to the digesters, which leads to longer digester retention times that may result in (a) greater solids destruction, and subsequently less solids hauling, and (b) greater gas production (if anaerobic digestion is in use). Thus, through a cascade effect, selective intermittent pumping of primary sludge can reduce the overall energy footprint of the treatment process through a number of avenues.

3.0 SECONDARY TREATMENT

3.1 Activated Sludge Processes

Following preliminary and primary treatment, the wastewater contaminants remaining consist of colloidal matter that is highly organic as well as a small amount of dissolved organic matter, nutrients, and dissolved inorganic solids. Colloidal and dissolved organic matter is amenable to biological secondary treatments. There are numerous methods used to accomplish biological treatment, but most can be categorized as either suspended-growth or fixed-film systems. Suspended-growth processes involve sustaining an active culture of microorganisms in suspension within a reactor, where air or oxygen is introduced to maintain biological activity. The demand for oxygen within a biological reactor can be met in various ways, but most aeration devices currently being used may be classified as either diffused, dispersed, or mechanical aeration systems.

Overall, aeration devices used for the activated sludge system represent the most significant consumers of energy within a wastewater treatment system. The ability of any type of equipment to dissolve oxygen in wastewater will depend on many factors, including diffuser device type, basin geometry, diffuser depth, turbulence, ambient air pressure, temperature, spacing and placement of the aeration devices, and diurnal variations in wastewater flow and organic load.

The most important energy decision for secondary treatment is made when the design is finalized. Once a system is installed, it is up to the operator to maintain the system at peak efficiency. Energy considerations for aeration systems, including the method for calculating energy use, are covered in greater depth in Chapter 8.

There are numerous configurations currently used at WWTPs for suspended-growth systems. The activated sludge systems operating today may be used for carbonaceous removal and nitrification, biological phosphorus removal, or complete biological nutrient removal systems. Carbonaceous and nitrification systems use aeration throughout the aeration basin(s). These facilities may be constructed as single-pass tanks or multiple-pass tanks. If this is an existing system, then it is up to the operating staff to maximize the energy efficiency of the system. This can be accomplished operationally by maintaining the proper dissolved oxygen within the basins. Having high dissolved oxygen concentrations within aeration tanks is a waste of energy. Suspended-growth systems should maintain dissolved oxygen concentrations between 0.5 to 2 mg/L. If the system uses blowers, then an operator should cut back on blowers or blower output. If the facility has coarse bubble diffusers, then a fine-bubble-diffuser system that is more efficient and uses less energy should be used.

If a facility has surface aerators, the submergence on the unit should be decreased. This will reduce the dissolved oxygen concentration as well as the amperage load on the motor. Reducing the electrical load will reduce electrical costs. If the liquid level of the basin cannot be adjusted, then VFDs should be installed on the aerators or the aerators should automatically start and stop based on time intervals. This mode of operation will depend on tank configuration, process considerations, and electrical demand charges. An analysis of the facility, process conditions, and utility charges must be investigated first to determine if energy savings can be realized as well as to determine that there will not be any detrimental effects on the process.

Biological phosphorus removal systems and biological nutrient removal (BNR) systems use both aerated and un-aerated zones to fulfill their treatment requirements. The un-aerated zones (anaerobic and anoxic zones) should contain efficient mixers that are driven by VFD units. Variable-frequency drive units allow the mixer to be turned down while maintaining proper mixing and saving energy. Biological nutrient removal processes also require internal recycling system to maintain proper nitrate levels in the anoxic zone. This internal recycle flowrate could be 300 to 400% of the forward flow, which represents a substantial quantity of water to pump. The pumps used for this purpose must be high-volume, low-head type pumps. In addition, the pumps must be on VFDs in order to return the correct quantity of flow to maintain the proper operating conditions while maximizing energy savings.

3.2 Dissolved Oxygen Operating Levels

The activated sludge process at a WWTP is one of the highest energy consumption areas of the treatment facility. Today's activated sludge systems use many different devices to supply and mix the contents of the aeration tanks. Air is supplied to aeration tanks by centrifugal-type blowers, positive-displacement blowers, high-speed turbine (HST) blowers, surface aerators, or a combination type system that uses a mechanical mixer and draft tube that is supplied air by a blower. Activated sludge aeration equipment uses large horsepower motors and multiple units to successfully meet the oxygen demands of the system. Adding excessive air into the aeration tanks will only result in a waste of energy. A general rule of thumb has always been to maintain a dissolved oxygen concentration in the aeration tanks of 2 mg/L. Once the carbonaceous demand and the nitrogeneous oxygen demand have been satisfied, maintaining any dissolved oxygen concentrations above 2 mg/L is excessive and a waste of energy. Many WWTPs today operate at low dissolved oxygen concentrations after they meet their carbonaceous and nitrogeneous oxygen demands. A number of treatment facilities maintain a dissolved oxygen concentration at the end of the aeration basin in the range of 0.5 to 2.0 mg/L.

Dissolved oxygen meters and blower control systems are more reliable today than they were a few years ago. The latest dissolved oxygen control systems can maintain tighter control ranges on the system, thus maintaining consistent energy consumption. There are several different dissolved oxygen and blower control modes of operation. The two common modes of operation are dissolved oxygen control only and dissolved oxygen control and blower discharge pressure control. The dissolved oxygen control mode only monitors dissolved oxygen control in the aeration basin and then sends a signal to increase or decrease the air to maintain a setpoint. The second system that uses dissolved oxygen control and blower discharge pressure maintains tighter blower controls by keeping the blower at its optimum operating point while still maintaining a preset dissolved oxygen concentration in the aeration tank.

In oxidation ditch systems, the dissolved oxygen concentration is usually varied by adjusting the submergence of the surface aerator. Automatic dissolved oxygen control systems for oxidation ditches measure the dissolved oxygen near the effluent weir. The dissolved oxygen meter transmits a signal through the plant's computer system to adjust the effluent gate upwards or "down-water" to maintain the preset dissolved oxygen concentration. Adjusting the effluent weir changes the mixed liquor level within the basin which, in turn, modifies the submergence on the aerator.

Maintaining a constant dissolved oxygen concentration in the aeration basin will save energy because the blower load can remain fairly constant and/or the blower can adjust within a smaller range, thus keeping energy use constant.

There are many diffused aeration systems that are used at WWTPs. Some plants use coarse-bubble diffusers, whereas a majority of facilities have converted to fine-bubble diffusers. Fine-bubble diffusers provide higher oxygen-transfer efficiencies as opposed to coarse-bubble diffusers. Installing fine-bubble diffusers into aeration basins will reduce the blower capacity needed to satisfy the process air demands. Less blower capacity will result in energy savings.

3.3 Secondary Clarification

Most WWTPs use secondary clarifiers for solids settling from an activated sludge process or humus from a tricking filter (fixed-film) process. Sequencing batch reactors (SBRs) settle solids in the same tank where the process treatment is occurring. The secondary clarifiers used in activated sludge and fixed-film processes contain collector mechanisms for gathering and removing the settled solids from the tanks. These collector mechanisms use small horsepower motors, typically in the range of 0.5 to 1.5 kW. Large, rectangular secondary clarifiers that use cable and drag-type collection mechanisms may use slightly higher horsepower motors in the range of 3 to 5 kW. Secondary clarifiers are not large energy consumers and, therefore, little to no energy savings can be found in this area.

3.4 Membrane Bioreactor Process

The membrane bioreactor process (MBR) incorporates the process features of a conventional secondary or advanced nutrient removal activated sludge process. Its unique feature is the use of membranes to separate solids from the mixed liquor instead of secondary clarifiers. The process eliminates some of the equipment and structures normally required for complete treatment. The use of membranes allows the mixed liquor to be concentrated much higher than in normal activated sludge reactors. Although high solids levels can upset clarifiers, they have limited impact on membranes. Aerobic reactors associated with MBRs can therefore be much smaller than conventional activated sludge reactors.

However, MBRs are generally considered high-energy-use processes. Unlike secondary clarifiers, which require little energy to operate, membranes have several features that require greater energy for the treatment plant as a whole:

- Immersed membranes act as fine-pore-size filters and require a driving head, or transmembrane pressure, to pass liquid through the media. Transmembrane pressures ranging from 1.5 m (4.9 ft) to 6.0 m (19.7 ft) are required to operate these microfiltration or ultrafiltration units. Whether these are static driving heads or suction from a pump, they add power requirements for the MBR.

- Accumulated solids on the membrane surface increase transmembrane pressures on membranes and reduce permeability and, in turn, the capacity of the system. Membranes require continuous or intermittent/cyclic air scouring to keep the membranes relatively free of caked solids and to maintain filtering capacity. Air scour systems usually consist of coarse bubble diffusers located at the base of the membrane cassettes..

- Higher aeration system requirements are due to lower alpha associated with high mixed liquor suspended solids.

- Higher recycle rates are required to move solids from the membrane tanks back to the aeration tanks.

- These systems have numerous control valves and vacuum systems to perform auxiliary functions.

- Backpulsing and chemical cleaning must be performed on the membranes regularly, often requiring additional pumping systems and chemical feed systems. Original designs had backpulsing of each membrane every 10 minutes, requiring numerous valve operations.

Further adding to energy requirements is the need for fine screening (to 1 to 2 mm) ahead of the bioreactor, which produces greater volumes of screenings and larger inlet works systems. Therefore, the benefits of MBRs (small footprint and super clarity permeate) must be tempered with higher energy requirements, which may be twice that of conventional activated sludge systems.

Membrane bioreactor suppliers and consulting engineers have been looking at energy-saving options to reduce the energy profile of these systems. If they can be made more energy efficient, they may be more suitable for large facilities.

Among the options investigated for energy savings are sequencing scour cycles to reduce the number of blower units in service, passive recirculation sludge return, pacing of sludge recycle with influent flows, reduction in the number of control valves required for operation, and elimination of active backpulse operation in favor of relax-mode operation.

In early designs, one membrane scour air blower was dedicated to a cassette tank, requiring every blower to operate continuously. One design change that can reduce scour aeration (and energy) requirements by 50 to 75% requires delivery of scour air in a common header to all membrane tanks, with controlled outlets to each tank. Instead of scouring each tank continuously, air is fed to tanks sequentially over a cycle of several minutes, allowing a single blower to serve the needs of all tanks. In one instance, this reduced operation from four blowers to a single blower operating at optimum efficiency.

3.5   Anoxic-Zone Mixing

Anoxic-zone mixing is typically accomplished by submerged mixers, slow-speed surface mixers, or minimal diffused-air addition. If variable-rate control of the mixers or blowers is available, there may be an opportunity to optimize mixing and minimize energy required. It is essential that the process goals of satisfying permit requirements be the priority in optimizing this process.

3.6   Fixed-Film Processes

Fixed-film processes consist of a fixed media on which microorganisms grow. The media can be any type of non-biodegradable material that will maintain its structure when exposed to water for a long period of time. Media commonly used are plastics, rock, slag, coal, and redwood. Fixed-film processes such as trickling filters typically require that the pretreated wastewater be pumped a second time in the overall WWTP process, with oxygen being supplied as the wastewater "trickles" down through the media. These units often require recirculation to provide both a minimum wetting rate as well as a more uniform wetting rate. Recirculation rates may be as much as three times the daily influent flow. Rotating biological contactors typically do not require repumping of the wastewater to the unit, but often use pumped recirculation. Rotating biological contactors use rotational energy for supplying oxygen rather than pumping. Fixed-film processes use less energy than suspended-growth processes, but do not achieve as high a degree of treatment. Additionally, fixed-film processes are more prone to odors. Once odor systems are installed, the relative energy efficiencies or at least overall operating costs of fixed-film processes begin to approach those of suspended-growth processes.

Most secondary treatment systems use a secondary clarifier for capture and recirculation of the biological solids created in the process. Secondary clarifiers are not large consumers of energy and typically use fractional horsepower drives.

Energy-use curves for various secondary treatment processes were developed by U.S. EPA (1978). These curves provide energy utilization rates (in kilowatts per year) for a WWTP flow capacity range of 0.04 to 4 m^3/s (1 to 100 mgd).

3.7   Online Instrumentation

Improving the operation of aeration systems is one of the best ways to reduce the energy costs for wastewater treatment. The amount of oxygen required by the activated sludge process varies as the flowrate and organic waste load change. A key ingredient in achieving effective and efficient treatment is supplying proper amounts of oxygen. Dissolved oxygen probes (sensors), in conjunction with online instrumentation through the use of supervisory control and data acquisition (SCADA) systems, perform the critical function of measuring dissolved oxygen levels in the aeration process so air flowrates can be controlled. Typically, oxygen requirements vary throughout the day by a factor of 5 to 7. Dissolved oxygen can be regulated by automatically controlling the aeration devices. Automatic control systems, which consist of probes, transmitters, and controllers, maintain preset dissolved oxygen concentrations in the aeration process. Air flowrates may be automatically adjusted by changing blower speed, adjusting blower inlet guide vanes, or operating control vanes.

The dissolved oxygen probe is cell-immersed in wastewater to measure dissolved oxygen. Dissolved oxygen probes measure oxygen quickly and continuously and eliminate the need to collect, transport, and analyze samples. The dissolved oxygen signal can then be connected to the SCADA system to provide real-time dissolved oxygen measurement information.

There are three technologies that are primarily used in wastewater applications:

1. *Galvanic*—These probes use an anode and a cathode immersed in an electrolyte behind a semi-permeable membrane. The membrane allows dissolved oxygen to pass into the measuring area. The amount of current the anode/cathode allows is proportional to the amount of dissolved oxygen present.
2. *Polarographic*—These probes use the same technology as the galvanic probes, but without the membrane.
3. *Optical*—These probes use a light source that pulses a certain amount/wavelength of light. The light pulses excite a fluorescent material. The amount of light emitted by the fluorescent material is proportional to the amount of dissolved oxygen in the solution.

There are positives and negatives associated with each technology, as well as various maintenance requirements that should be accounted for in any application.

A key to controlling the activated sludge process (and energy use) is matching oxygen supply to oxygen demand. Typically, as wastewater flow and oxygen demand increase in the morning based on the diurnal flow of a WWTP, more aeration is needed.

Conversely, as flow decreases during the night, the air supply should be reduced. In a diffused-air system equipped with centrifugal blowers, dissolved oxygen can be controlled by regulating blower air flowrates. Air flowrates can be changed to meet process oxygen requirements by adjusting the position of the inlet guide vanes. It must be noted that the adjustment of air flow depends on the type of blower, and guide vanes are not appropriate for all blowers. Other methods of control include throttling the blower inlet or adjusting blower speed, if available. Air flowrate is regulated based on oxygen levels as measured by the dissolved oxygen probes. When lower air flowrates are required, the blower power draw is reduced, thereby saving electricity. Dissolved oxygen control could reduce energy by as little as 10% or as much as 35% based on improvements over manual control.

The control method used to fluctuate the air flowrate varies based on the mechanical aeration device. Typically, the primary control input for all modes of control is dissolved oxygen; however, it can include mixed liquor solids, return solids, and influent flow. Control methods vary from modulating inlet guide vanes or butterfly valves, modulating isolation valves to aeration basins, or varying blower speed through the use of VFDs (see Figure 7.3).

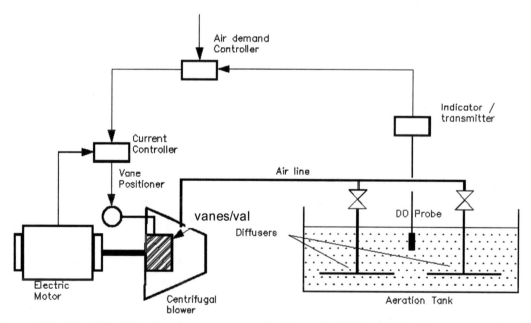

FIGURE 7.3   Simplified dissolved oxygen control schematic (EPRI, 1996).

Control can be done by a master controller, typically a programmable logic controller, remote telemetry unit, or distributed control system. These master controllers typically have enough intelligence to monitor and control all the devices associated with a typical control mode for the aeration process.

4.0 DISINFECTION

Disinfection of final effluent inactivates or destroys most of the microorganisms remaining in the wastewater after treatment, including pathogens. Presently, the most common method of disinfection is through the addition of chlorine gas solutions or hypochlorite solutions. Due to concerns over the handling of chlorine gas and/or formation of disinfection byproducts with the use of chlorine, UV radiation is becoming an increasingly more common wastewater disinfection technology. Ozone disinfection has also received some interest, but is relatively uncommon in wastewater treatment. Both UV and ozone disinfection require considerably greater quantities of primary energy than chlorination.

4.1 Chlorination/Dechlorination

The methods and equipment used for chlorination of wastewater effluent are similar to chlorination methods described in Chapter 6 for water treatment. Disinfection with chlorine gas and bulk hypochlorite requires only a minor amount of electrical energy due to consumption by:

- Evaporator heaters (chlorine gas),
- Pumping of dilution water, and
- Metering and pumping of chlorine/hypochlorite solution.

Electrical consumption at a treatment facility is higher for on-site generation of hypochlorite, which requires approximately 20 MJ (megajoules) (2.5 kWh) of electricity to produce 1 kg of equivalent chlorine (2.5 kWh/lb) as a 0.8% sodium hypochlorite solution. The major power consumption in hypo-chlorite generation systems is due to conversion of brine solution to caustic and chlorine gas streams in electrolytic cells, which contributes to more than 50% of the overall production cost of hypochlorite (Casson and Bess, 2003). Off-site generation of chlorine chemicals consume energy even though energy consumption is generally not directly attributed to the treatment facility.

Dechlorination of wastewater (after chlorine disinfection) is typically achieved by the addition of sulfur dioxide gas or sulfite salt solution. Sulfur dioxide gas feeders (sulfonators) and sulfite salt solution feed systems are similar in design to chlorine gas feed and hypochlorite feed systems, respectively (WEF and American Society of Civil Engineers, 2009). As with chlorination systems, electrical consumption of dechlorination systems is relatively low.

A 2006 New York State Energy Research and Development Authority (NYSERDA) (Albany, New York) study evaluated energy use at seven WWTPs in New York that use chlorination/dechlorination, ranging in capacity from 3.5 to 135 mgd (0.15 to 5.9 m³/s). The energy consumption of disinfection processes was approximately 0.2% of the total energy consumption of the wet treatment processes in the seven plants, indicating that energy use by chlorination/dechlorination processes is insignificant (Malcolm Pirnie, 2006).

4.2 Ultraviolet Disinfection

Ultraviolet radiation is an effective disinfection process for many bacteria, protozoa, and viruses present in secondary wastewater effluent. Ultraviolet disinfection systems for wastewater are similar in many respects to the UV systems for water disinfection described in Chapter 6. A significant difference is that ultraviolet transmittance (UVT) is typically lower and suspended solids are generally higher in wastewater systems than in water systems. Ultraviolet transmittances in secondary wastewater typically range from 45 to 70%, whereas UVTs in drinking and reuse water treatment typically range from 70 to 95%. Lowering UVT increases energy consumption because UV light will not propagate as far in low UVT waters, requiring UV systems with more lamps and a narrower lamp spacing. Depending on the level of disinfection required, an upstream solids removal process (such as granular media filtration or membrane bioreactor) may be required to reduce the level of solids that contain shielded microbes. The upstream solids removal process increases plant energy use.

Wastewater UV disinfection systems typically use open-channel reactors with either low-pressure, low-pressure high-output, or medium-pressure mercury lamps. Lamp types differ significantly by the following:

- Total power consumption per lamp,
- Conversion efficiency from electrical energy to germicidal UV output, and
- Total number of lamps required.

Low-pressure high-output (LPHO) UV systems typically require less total energy to deliver the same UV dose than low- and medium-pressure systems (American Water Works Association Research Foundation, 2001; U.S. EPA, 2006). However, LPHO and low-pressure systems require a significantly higher number of lamps and higher space requirements than medium-pressure systems. For this reason, medium-pressure systems are generally better suited for high-flow systems, stormwater over-flows, or space-limited sites (Metcalf & Eddy, 2007). Similar to drinking water systems, wastewater UV systems use ballasts to control power to UV lamps (either electronic or electromagnetic). Ballasts allow turndown of lamp energy to as low as 30% of maximum lamp output depending on the ballast type and design. Electronic ballasts are approximately 10% more energy efficient than electromagnetic ballasts (Metcalf & Eddy, 2007; U.S. EPA, 2006).

Power consumption in wastewater UV systems can be regulated by a "dose-pacing" strategy that controls the number of lamps/banks in operation and/or the relative output of the lamps. With this approach, the number of banks of lamps in operations is typically controlled by the flowrate, whereas the relative lamp output is typically controlled by the measured UVT (U.S. EPA, 2006). Dose-pacing control can be implemented in a number of different methods, which vary by system and UV manufacturer.

A 2004 NYSERDA study evaluated power consumption for chlorination/dechlorination and three UV-disinfection alternatives (low pressure, LPHO, and medium pressure) for a 0.8-m³/s (18-mgd) WWTP in New York State (URS Corporation, 2004). Table 7.1 summarizes power use for each alternative to meet a 200 fecal coliform units (CFU) /mL standard, which shows that power costs for UV disinfection were significantly higher than for chlorination/dechlorination. Low-pressure high-output had the lowest projected power consumption of the three UV technologies considered.

TABLE 7.1 Estimated power consumption for disinfection alternatives for 18-mgd WWTP in the NYSERDA study (URS Corporation, 2004)

	Alternative 1	Alternative 2	Alternative 3	Alternative 4
Disinfection Process	**Chlorination/ Dechlorination**	**UV – LP**	**UV-LPHO**	**UV-MP**
Annual Power Consumption (kWh/d)	144	1440	1080	4560

5.0   ADVANCED WASTEWATER TREATMENT

5.1   Granular Media Filtration

Granular media filtration is a traditional method used in wastewater treatment and represents an advanced secondary or advanced wastewater treatment process for removal of suspended solids. Historically, fine-mesh (micro) screens were used for granular media filtration, and there are evolving technologies that use cloth and other alternative medias.

The filtration operation uses energy to overcome head loss through the filter media. Energy is also used to backwash the media to remove accumulated solids. There are two general categories of filters: those in which the water passes through the media using gravity head and those in which water passes through media in an enclosed pressure vessel.

In filtration operations that use gravity flow, an allowance for head loss through the filter media, which may range from 1 ft to more than 5 ft, has been built into the hydraulic profile for the plant. Pumps somewhere in the system must overcome the entire designated head loss regardless of the actual filter head loss. Flow through the filter is either at a constant flow, with an increasing head loss as the media collects solids, or at a fairly constant head with a declining flow.

In totally enclosed pressure filters, head loss builds as the media becomes fouled. Although head loss varies with the buildup of debris, overall losses may be as severe as or worse than with open filters. For either gravity or pressure filters, backwashing removes the accumulated solids and restores the head loss back toward the original condition. Frequent backwashing will reduce the head loss during filtration, but will result in shorter filter runs, lower production rates, and increased energy for backwashing. Extending the duration between backwashing events will tend to increase head loss through the filters. There are a number of other factors that affect the frequency of backwashing, such as the load of solids in the water to be filtered and the maximum length that a filter should run between backwashes. The operator should assess these variables to determine the lowest cost to operate.

Ancillary processes such as filter-aid chemical pumping, polymer-feed systems, and surface air-scouring and wash systems also use energy for mixing, pumping, and air-compressor power, respectively. Although these tend to be small demands, they should be part of the overall energy management plan for a filtration facility.

Similar considerations will apply to other granular media processes such as activated carbon adsorption and ion exchange. Backwashing can be undesirable with these processes as it mixes the bed, reducing the overall time to break through. This may be acceptable if there are multiple units operating in series.

Submerged disc filtration is becoming increasingly popular for tertiary filtration applications. Disc filtration offers potential energy savings due to the lower head loss experienced across the filter.

5.2   Activated Carbon Adsorption

Energy is consumed in the activated carbon adsorption process to overcome head loss in the contacting mode, which is similar to the built-in head losses with granular media filtration described previously. The spent carbon requires fuel for thermal regeneration to operate a pyrolysis furnace or to furnish steam. Some carbon systems use chemicals such as caustic soda for regeneration. Electricity is required to operate the carbon transfer system.

5.3   Chemical Treatment

Energy required for chemical treatment can be divided into four parts: chemical feed, rapid mix, flocculation, and sedimentation. Sludge pumping and chemical regeneration such as lime recalcination can consume large amounts of energy. In chemical treatment processes, the largest amount of energy use is the secondary energy required in the manufacture of the chemicals.

5.4   Nutrient Removal Processes

Control of nutrients in WWTP effluents is becoming increasingly important as more stringent discharge requirements are imposed. In freshwater receiving waters, phosphorus has typically been considered the limiting nutrient in the harmful phenomenon of eutrophication, which is primarily manifested by algal blooms that occur during summertime. However, for some estuarine and other waterbodies, control of both phosphorus and nitrogen may be necessary to prevent deterioration of water quality.

The two possible techniques used for phosphorus removal are traditional chemical phosphorus removal (CPR) and biological phosphorus removal (BPR). Chemical phosphorus removal is achieved via precipitation of phosphorus in the soluble form by adding metal salts such as aluminum sulfate (alum) and ferric chloride. Although effective, this technique can be costly due to high chemical costs as well as increased sludge treatment and disposal costs. Figure 7.4 presents the required metal dosage of

| Phosphorus | Mole ratio, Al:P | |
Reduction, %	Range	Typical
75	1.25:1–1.5:1	1.4:1
85	1.6:1–1.9:1	1.7:1
95	2.1:1–2.6:1	2.3:1

Developed in part from U.S. EPA (1976)

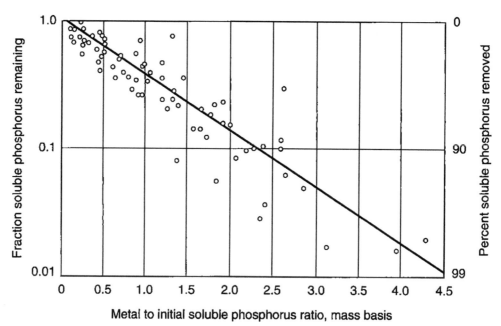

FIGURE 7.4   Metal dose requirement for phosphorus precipitation in wastewater using alum (table) and ferric chloride (chart) (Sedlak, 1991).

aluminum and iron to achieve various levels of phosphorus removal. As the figure shows, the metal dosage increases as the target residual of phosphorus approaches a low level, which will impact the cost of chemicals significantly. The multiple-point metal salt addition approach can be applied to reduce the required dosage.

Biological phosphorus removal provides a cost-effective means for the removal of phosphorus from wastewater. Biological phosphorus removal processes use a non-aerated zone that selects for the growth of specific organisms called *polyphosphate-accumulating organisms* (PAOs). Polyphosphate-accumulating organisms are capable of storing excessive amounts of phosphorus in their body cells (5 to 7% of body mass) under certain environmental conditions. In the absence of oxygen in the nonaerated zone, PAOs take up the readily biodegradable organic carbon in the wastewater and release phosphorus. Then, in the aerated zone, they excessively take up phosphorus and store it in their body cells, which then will be wasted in the sludge stream. From the standpoint of nutrient recovery, BPR processes are better suited for phosphorus and nitrogen recovery than CPR processes.

Conventional nitrogen removal is accomplished through biological nitrification and denitrification. Biological conversion of ammonia to nitrate (nitrification) requires oxygen to be supplied at 4.57 times the ammonia-nitrogen loading, which is a significant energy-consuming process. Additionally, because nitrification only occurs with low food-to-microorganism ratios, more energy is required to satisfy endogenous respiration demands. (See Chapter 8 for a more thorough discussion of aeration energy requirements.) During nitrification, alkalinity is consumed, where 7200 g of alkalinity as calcium carbonate ($CaCO_3$) is consumed per kilogram of ammonia-nitrogen oxidized (7.2 lb/lb). Alkalinity addition is required if the source in the wastewater is not sufficient.

Conversion of ammonia to nitrate is often sufficient for many discharge permits. However, in some instances, total nitrogen removal is the requisite. Nitrogen can be removed via air stripping of ammonia from effluents with pHs that have been elevated above 11 with alkali chemicals. This practice is relatively inefficient and temperature-sensitive, requires considerable amounts of chemicals (usually lime and neutralizing acid), is expensive, and adds offending ammonia to the air. As such, air stripping is not commonly seen in municipal WWTPs. An alternative process is biological denitrification following a nitrification process. However, biological denitrification requires an organic carbon source. If done as a separate process following secondary treatment, methanol is often used as the carbon source.

A preferred but not totally efficient method for denitrification is the Modified Ludzack-Ettinger (MLE) process that uses an aeration system with an anoxic zone partitioned off from the aerobic portion of the system. More complex processes are typically a modification to the Modified Ludzack-Ettinger process. The anoxic zone is located at the beginning of the aeration tank and uses mixers of some type rather than aerators or air diffusers. Nitrified wastewater from the rear of the aeration tank is recirculated to the anoxic zone, in which the nitrate is used as a source of oxygen, and the incoming BOD is used as the carbon source. As the nitrate gives up its oxygen, it breaks down to nitrogen gas and is released to the atmosphere.

The nitrate actually gives back some of the oxygen that was used in its formation in the nitrification step. Whereas it takes 4600 g of oxygen/kilogram of ammonia-nitrogen (4.6 lb/lb) to produce nitrate, the nitrate only gives back 2800 g of oxygen/kilogram of nitrate-nitrogen (2.8 lb/lb) (the additional oxygen that went to forming water in the oxidation of ammonia cannot be recovered). Additional amounts of energy are required for recirculation and for mixing the contents in the anoxic zone.

To be effective, the process requires high recirculation rates. Through a mass balance calculation, and assuming 100% utilization of nitrate in the anoxic zone, it can be shown that the efficiency of denitrification is related to the recirculation ratio in the following manner:

$$\text{Maximum efficiency of denitrification} = 1 - \frac{1}{N+1} \qquad (7.1)$$

Where

N = the number of recycles (i.e., the sum of all recycles that include nitrate) relative to the influent flow

For example, if the influent flow is 0.04 m³/s (1 mgd), the recirculation is 0.09 m³/s (2 mgd), and the number of recirculations is two, then the maximum efficiency of denitrification is 67%. This means that 33% of the nitrate would appear in the effluent and 67% would break down in the anoxic zone, yielding its oxygen and exiting the system as nitrogen gas. To attain 90% nitrogen removal would require nine recirculations; to attain 95% removal would require 19 recirculations. When high removal rates of nitrogen are required, it does not make sense to use the high recirculation rates needed. In addition, at high recirculation rates inhibition for denitrification becomes more problematic due to intrusion of dissolved oxygen into the anoxic zone. Consequently, a separate denitrification stage following secondary treatment will be required. Addition of primary effluent to the final stage of treatment can be done, but it brings up philosophical questions relating to the purification of the wastewater. More commonly, methanol is used as the carbon source, which can become an expensive process and generate additional sludge. However, several carbon source alternatives such as acetic acid, ethanol, glycerol, and other proprietary formulated products are being investigated at pilot- and full-scale applications in the United States to reduce cost and enhance performance. The general formula for the carbon requirement for denitrification can be expressed in the following manner (Metcalf & Eddy, 2007):

$$\text{Pound of CODbs oxidized/lb of nitrate-N reduced} = \frac{2.86}{1 - Y_n} \qquad (7.2)$$

Where

  CODbs = biodegradable soluble chemical oxygen demand

   Y_n = net biomass yield, lb CODvss/lb CODbs · d

 Alkalinity is reduced during the denitrification step, which reduces the need for alkalinity addition required in the nitrification step; in addition, 3600 g of alkalinity as $CaCO_3$ is produced per kilogram of nitrate-nitrogen reduced (3.6 lb/lb).

5.5 Side-Stream Nitrogen Removal Processes

Side-stream treatment processes for nitrogen removal offer cost-effective means to deal with the nitrogen-rich recycle streams generated from dewatering anaerobically digested sludges (centrate). Sequestering and removal of nitrogen loadings from centrate has been conducted in Northern Europe since the late 1980s and in North America since the early 1990s. Physicochemical processes (e.g., hot air and steam stripping, struvite precipitation) were the first to be developed and deployed. However, due to the cost-effectiveness of biological side-stream treatment technologies, several processes were developed in Northern Europe and North America in the early 1990s that are being continuously refined to this day. Centrate recycle flow typically contributes less than 1% of the wastewater flow to the secondary treatment process, but contributes approximately 15 to 25% of the total nitrogen load. A separate side-stream centrate treatment system can provide a cost-effective means for reducing the nitrogen load to the secondary treatment process via reduction in both the required bioreactor volume (i.e., smaller centrate reactor) and the operational cost. The volume reduction benefit is due to operating at a higher nitrification/denitrification rate (kg-N-removed/m³/day) than the secondary nitrification process. The higher rates are maintained in the side-stream process due to the warm environment (typically 30 to 35 °C), allowing for higher nitrification and denitrification rates. The potential reduction in operational cost is dependent on the type of treatment process selected for centrate treatment. Any biological centrate treatment process can be classified as one of the following three types: nitrification/denitrification, nitritation/denitritation, and partial nitrification/anaerobic ammonia oxidation (ANAMMOX), also known as *deammonification*.

 Figure 7.5 illustrates the distinction between the three categories of nitrogen removal. As shown in the nitrogen cycle, ammonium is oxidized in two primary steps, nitritation (ammonium to nitrite) and nitratation (nitrite to nitrate) by aerobic autotrophic bacteria ("nitrifiers"). Collectively, the two steps are referred to as *nitrification*. During this process, oxygen is consumed and alkalinity is destroyed via the production of acidity. Similarly, the denitrification process consists of two primary steps: reduction of nitrate to nitrite (denitratation) and nitrite to nitrogen gas (denitritation). These steps are conducted under anoxic conditions and are largely performed by heterotrophic bacteria capable of using nitrite and nitrate as the electron acceptor in the absence of dissolved oxygen. These two denitrification steps require an organic carbon source such as methanol (external source) or constituents in the raw or primary effluent that compose the readily biodegradable fraction of BOD. As a product of the denitrification process, bicarbonate alkalinity is generated. Collectively, the entire nitrogen process is called *nitrification/denitrification.*

In the second process category, referred to as *nitritation/denitritation,* the oxidation of ammonium is stopped at nitrite (nitritation). In the subsequent anoxic step, heterotrophic bacteria reduce the nitrite to nitrogen gas using an organic carbon source (denitritation). The advantage of this type of nitrogen removal process is illustrated in Figure 7.6. In terms of the relative demand for resources (oxygen and organic carbon source), performing only nitritation reduces the oxygen demand by 25% because nitrite is not being further oxidized to nitrate. During denitritation, 40% less degradable carbon or BOD is required to produce nitrogen gas compared to the total carbon demand required to reduce nitrate to nitrite and nitrite to nitrogen gas.

In the third process category, referred to as *partial nitritation/anaerobic ammonium oxidation* (collectively called *deammonification*), a unique group of anaerobic autotrophic bacteria (ANAMMOX bacteria) are capable of consuming ammonium and nitrite simultaneously to produce nitrogen gas and a small amount of nitrate

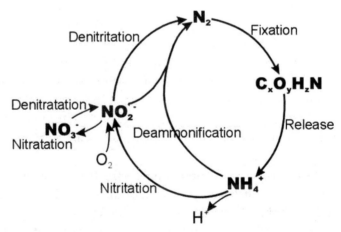

FIGURE 7.5   Nitrogen cycle in wastewater treatment processes (Wett, 2007).

(nitrate product is not shown in Figure 7.5). Because the ANAMMOX bacteria do not require an organic carbon source for growth, this drastically reduces the carbon or BOD requirement for nitrogen removal, as indicated in Figure 7.6; the only carbon that is needed is for the further reduction of nitrate produced by the reaction. To provide the ANAMMOX bacteria with nitrite and ammonium, the nitritation reaction is not performed to completion (i.e., only partial conversion of ammonium to nitrite or only a portion of the centrate is processed through the nitritation step and then blended with the untreated centrate to provide both ammonium and nitrite to the ANAMMOX bacteria.) In either instance, the oxygen demand is further reduced, as indicated in Figure 7.6, because only a portion of the ammonium is being aerobically oxidized to nitrite. Operating cost certainly favors the deammonification processes.

At the Strass WWTP in Austria, the site of the first DEMON® SBR process, the power savings anticipated by the implementation of deammonification were realized. Previously, power consumption in the nitritation/denitritation operating mode was measured at approximately 10.4 MJ/kg of nitrogen removed (2.9 kWh/kg) (aeration, mixing, and pumping). After the final loading target was achieved, six months after seeding the reactor with ANAMMOX-enriched biomass, the power consumption in the

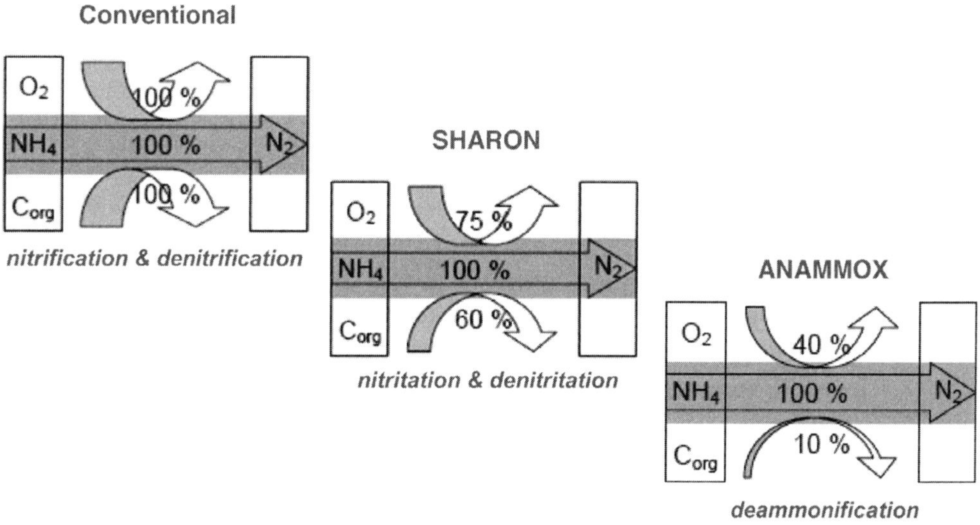

FIGURE 7.6   Relative demand for resources for the three primary types of nitrogen removal processes (Wett, 2006).

process decreased to approximately 4.2 MJ/kg of nitrogen removed (1.16 kWh/kg). In comparison, a power consumption rate of approximately 23.4 MJ/kg of nitrogen removed (6.5 kWh/kg) was measured in the Strass secondary treatment process.

5.6   Post-Aeration

Post-aeration, which is usually the last treatment process in WWTPs, is needed at some plants to assure an adequate level of dissolved oxygen in the wastewater discharged from the plant. Post-aeration may be provided by diffused aeration, mechanical aeration, air ejection, or by cascades. Energy consumption is primarily due to the aeration blowers or mechanical aerators used to provide aeration.

Diffused aeration is the most common technology used in post-aeration. In smaller plants, course-bubble diffusers are used because of the nominal amount of oxygen that must be transferred. Larger plants will likely use membrane or fine-pore diffusers. The dissolved oxygen should be monitored and the output of the aeration blowers controlled to provide the amount of dissolved oxygen needed.

Floating mechanical aerators have also been used. It will likely be most efficient to use this type of device to provide additional oxygen during those times that experience shows additional aeration is needed.

Air ejectors may be useful where the plant does not have the space for a post-aeration basin or the drop for a cascade. Treated effluent is pumped through the ejector, which aspirates air or concentrated oxygen into the water. Energy is consumed in the head loss through the ejector, which is a function of the ejector's geometry and the rate of flow. Additional operating energy will be required if a dedicated pumping system needs to be installed to operate the ejector.

Cascades require no direct energy input, using the natural drop that is available at the outfall. However, if this drop is reserved for the cascade at the expense of the need for additional pumping, then the cost of pumping should be considered in the evaluation of this method of post-aeration. Although cascades are an effective and energy efficient means of post aeration, the rate of aeration and resulting dissolved oxygen concentration in the effluent cannot be accurately controlled. The choice of post-aeration process may impact the hydraulic profile of the plant, and thus the overall energy requirements of the facility.

6.0   MISCELLANEOUS ENERGY USES

There are many auxiliary processes in a WWTP that require energy. Examples are seal water, service water, auxiliary service compressed air, instrument air, hoisting cranes, sump pumps, and, sometimes, potable water. Some minor energy uses include instrumentation, electric-operated valves, gas pilots for boilers, digester gas flares, unit heaters, and portable steam units. Additionally, an emergency power supply may be available in the form of engine generators.

Wastewater treatment plants have buildings for housing equipment and personnel for facility operation, maintenance, laboratory, and administration. All of these buildings use energy for space heating during colder weather in cold climates and air-conditioning during warmer weather. Energy is also used for air-handling units for ventilation, odor control units, and for lighting work areas, outside areas, and interconnecting gallery areas.

7.0   REFERENCES

American Water Works Association Research Foundation (2001) *Practical Aspects of UV Disinfection;* American Water Works Association Research Foundation: Denver, Colorado.

Casson, L.; Bess, J., Jr. (2003) *Conversion to On-Site Sodium Hypochlorite Generation: Water and Wastewater Applications;* Lewis Publishers: Boca Raton, Florida.

Electric Power Research Institute (1996) *Improving Operation of Aeration Systems Using DO Probes;* Electric Power Research Institute: Palo Alto, California.

Malcolm Pirnie (2006) *Municipal Wastewater Treatment Plant Energy Evaluation Summary Report;* New York State Energy Research and Development Authority: Albany, New York.

Metcalf & Eddy (2004) *Pumping Systems Training Manual;* McGraw-Hill: New York.

Metcalf & Eddy (2007) *Water Reuse: Issues, Technologies and Applications;* McGraw-Hill: New York.

National Fire Protection Association (2008) *Standard for Fire Protection in Wastewater Treatment and Collection Facilities;* NFPA-820; National Fire Protection Association: Quincy, Massachusetts.

URS Corporation (2004) *Evaluation of Ultraviolet (UV) Radiation Disinfection Technologies for Wastewater Treatment Plant Effluent;* New York State Energy and Research Development Authority: Albany, New York.

U.S. Environmental Protection Agency (1978) *Energy Conservation in Municipal Wastewater Treatment;* MDC-32; Washington, D.C.

U.S. Environmental Protection Agency (1979) *Process Design Manual for Sludge Treatment and Disposal*; EPA-625-1-79; Washington, D.C.

U.S. Environmental Protection Agency (2006) *Ultraviolet Disinfection Guidance Manual for the Final Long Term 2 Enhanced Surface Water Treatment Rule*; EPA-815-R-06-007; Washington, D.C.

U.S. Environmental Protection Agency (1989) *Design Manual: Fine Pore Aeration Systems*; EPA-625/1-89/023; Washington, D.C.

Water Environment Federation; American Society of Civil Engineers (2009) *Design of Municipal Wastewater Treatment Plants*, 5th ed.; Manual of Practice No. 8, ASCE Manual of Practice and Report on Engineering No. 76; Water Environment Federation: Alexandria, Virginia.

Water Pollution Control Federation (1983) *Nutrient Control*; Manual of Practice No. FD-7; Water Pollution Control Federation: Washington, D.C.

8.0   SUGGESTED READINGS

University of Florida (1986) *Operations and Training Manual on Energy Efficiency in Water and Wastewater Treatment Plants*; University of Florida, Center for Training, Research and Education for Environmental Occupations: Gainesville, Florida.

U.S. Environmental Protection Agency (1985) *Handbook—Estimating Sludge Management Costs*; EPA-625/6-85; Washington, D.C.

Chapter 8

Aeration Systems

(continued)

191

Electricity for aeration systems represents a substantial portion of the total energy use for most secondary and advanced wastewater treatment plants (WWTPs). It is therefore logical to focus initial efforts to achieve energy conservation on the aeration system. An aeration system is a collection of devices typically supplied by different manufacturers that, when combined, deliver oxygen to wastewater. A common aeration system consists of a blower, piping, valves, diffusers, and controls. Other methods of aeration include surface aerators, brush aerators, bio-towers, and other systems described in this chapter.

1.0   DETERMINING OXYGEN REQUIREMENTS

Energy conservation involves measurement of existing conditions to identify methods for reducing energy consumption. It also requires being able to measure post-improvement performance. Because aeration is a significant energy-consuming process and is used to meet oxygen requirements, it is important to estimate oxygen requirements as accurately as possible. It is also useful to distinguish between system design capacity and operating capacity.

When determining oxygen requirements for a new activated sludge system, in addition to several rational approaches, design engineers also apply a safety factor to ensure that the designed system has an adequate capacity to supply of oxygen for peak load conditions. Operators may use mixed liquor dissolved oxygen or another operating parameter as a guide to controlling oxygen supply to the mixed liquor suspended solids (MLSS). Operators may also use laboratory techniques to determine oxygen uptake rates, and use real time measurements to determine how much of the supplied oxygen is taken up as the air moves through MLSS.

A number of system and operational factors can affect the overall efficiency of oxygen transfer and, therefore, efficiency of energy use. Examples of commonly encountered conditions include the number of aeration basins in service, air rate per diffuser, presence of nitrification when not required, food-to-microorganism (F/M) ratio, aerator submergence, and air filter cleanliness.

One way to establish a "baseline" condition is to determine the amount of oxygen needed to remove the organic material from the wastewater being treated. Oxygen is required in a WWTP for the oxidation of organic matter, expressed as either carbonaceous biochemical oxygen demand or chemical oxygen demand (COD), and oxidation of ammonia. Organic matter is oxidized by organisms known as *heterotrophs.* Ammonia oxidation, a process known as *nitrification,* is carried out by organisms known as *nitrifiers.* Nitrification occurs when conditions in the WWTP enable the growth of nitrifiers.

In systems designed for biological nitrogen removal, nitrification is followed by denitrification, a process by which nitrites and nitrates are reduced to molecular nitrogen and released into the atmosphere. For denitrification to occur an electron donor (usually included in the measurement for COD) must be available and oxygen must be absent. If influent BOD can be utilized to supply the electron donor, the overall oxygen requirements of a WWTP can be reduced.

A part of COD entering a WWTP is converted into new biomass, which is harvested as waste activated sludge. The remaining COD is oxidized to carbon dioxide, which requires oxygen. The amount of biomass generated is based on the yield coefficient. In a conventional secondary system, the total biomass generated (yield) can be calculated as follows:

$$X = Y_{obs} (S_o - S_e) \tag{8.1}$$

Where

X = biomass generated, g/d;
Y_{obs} = observed yield, g biomass/g COD;
S_o = influent COD, g/m^3; and
S_e = effluent COD, g/m^3.

Observed yield, in turn, is a function of the solids retention time of the system and the decay coefficient, as follows:

$$Y_{obs} = \frac{Y_T}{1+(k_d \times SRT)} \tag{8.2}$$

Where

Y_T = theoretical yield coefficient, g biomass/g COD;

K_d = decay coefficient, g biomass destroyed/g biomass.d; and

SRT = solids retention time.

Biomass can be represented as $C_5H_7O_2N$ (Hoover and Porges, 1952). The theoretical COD of biomass can be calculated based on stoichiometry, by writing the chemical reaction of its complete oxidation, as follows:

$$C_5H_7O_2N + 5\,O_2 \rightarrow 5\,CO_2 + 2\,H_2O + NH_3 \tag{8.3}$$

Thus, complete oxidation of 113-g biomass requires 160 g of oxygen, which translates to 1.42 g of oxygen per gram of biomass. The COD incorporated into biomass does not exert an oxygen demand in the system. The oxygen requirement of the system for COD reduction can therefore be written as follows:

$$\text{Oxygen requirement for COD reduction} = S_o - [S_e + \{1.42Y_{obs}(S_o - S_e)\}] \tag{8.4}$$

Like COD, a part of the influent nitrogen is incorporated into biomass. Considering the chemical composition $C_5H_7O_2N$, biomass comprises approximately 12.3% nitrogen. Nitrogen that remains after incorporation into biomass is oxidized by nitrifiers. In the process, some nitrogen is incorporated into nitrifier biomass. Nitrifier biomass production can be estimated in a manner similar to the heterotroph biomass production, where all relevant kinetic coefficients pertain to nitrifiers.

Influent nitrogen comprises mainly free ammonia (NH_4^+-N) or organic nitrogen. Together, they are referred to as total Kjeldahl nitrogen (TKN). Almost all of the influent organic nitrogen is converted to NH_4^+-N and becomes available to biomass for new cell growth or nitrification. Complete nitrification occurs as follows:

$$NH_4^+ + 2\,O_2 \rightarrow NO_3^- + 2\,H^+ + H_2O \tag{8.5}$$

Based on the aforementioned reaction, 2 moles of oxygen are required to oxidize 1 mole of nitrogen to nitrate, which is equivalent to 4.57 g O_2 per g NH_4^+-N oxidized. Two equivalents of H^+ are produced in the process, which, in turn, reacts with two equivalents of bicarbonate in the wastewater. As a result, 7.14 g of alkalinity (as $CaCO_3$) are destroyed per g NH_4^+-N oxidized.

Considering a yield of 0.17 g of nitrifying bacteria, the overall nitrification reaction can be summarized as follows (Gujer and Jenkins, 1974):

$$1.02\ NH_4^+ + 1.89\ O_2 + 2.02\ HCO_3^- \rightarrow 0.021\ C_5H_7O_2N + $$
$$1.06\ H_2O + 1.92\ H_2CO_3 + 1.00\ NO_3^- \tag{8.6}$$

The oxygen requirement and alkalinity consumption in nitrification exhibit little change, even after considering biosynthesis, because of low bacterial mass yield. The oxygen requirement decreases to 4.3 g O_2 per g NH_4^+-N oxidized, whereas alkalinity consumption increases to 7.2 g as $CaCO_3$ per g NH_4^+-N oxidized. In design, nitrogen incorporation into biomass is typically ignored, and 4.57 g O_2 are provided per g NH_4^+-N oxidized.

1.1 Impact of Process Configuration, Nitrification, Denitrification, and Phosphorus Removal

Biological nutrient removal (BNR) systems have aerobic zones for nitrification and phosphorus uptake, anoxic zones for denitrification, and anaerobic zones to encourage the growth of phosphate-accumulating organisms (PAOs), which accomplish enhanced biological phosphorus removal. Phosphate-accumulating organisms sequester organic material under anaerobic conditions and store it for subsequent oxidation under anoxic or aerobic conditions. When mixed liquor enters aerobic zones of a BNR process after having passed through anaerobic and anoxic zones, most of the soluble organic material, which tends to decrease alpha (the ratio of the oxygen transfer coefficients in wastewater versus clean water), has already been removed. Further, as the organic material is already within the PAO cells, it is degraded rapidly, resulting in high oxygen uptake rates, which tend to decrease the bulk dissolved oxygen concentration. Thus, the driving force for oxygen transfer is increased. Oxygen transfer efficiency (OTE) is typically higher in BNR systems compared to conventional activated sludge systems (Mahendraker, Mavinic, Rabinowitz, and Hall, 2005; Mahendraker, Mavinic, and Rabinowitz, 2005; Rosso et al., 2008).

In the aerobic zones of BNR systems, ammonia is oxidized to nitrite and nitrate. The energy spent in providing oxygen for the oxidation of ammonia is partially recovered as the oxidized nitrogen is used as an electron acceptor in the anoxic zones, thus reducing the overall oxygen requirement. Each kilogram of NO_3-N reduced to nitrogen gas consumes 2.86 kg of COD, thus reducing the oxygen requirement by that amount.

In BNR processes, several process configurations take advantage of the residual dissolved oxygen in the return activated sludge (RAS) without affecting nitrogen and phosphorus removal capacity of the process and reducing the total aeration requirements of the process. The actual design is a function of influent wastewater and target effluent quality. Return activated sludge needs to be designed in such a way so as to increase denitrification rates within the anoxic reactor and eliminate ingress of nitrates or oxygen in the anaerobic reactor to maintain PAOs. In the Modified University of Cape Town (MUCT) process at least two options of using residual dissolved oxygen exist: (1) RAS feed to the aerobic reactor and (2) RAS feed to anoxic reactor. In the Modified Johannesburg (MJB) process, the RAS loop should have a small de-oxygenation and denitrification zone followed by RAS split flow to the anaerobic (25%) and anoxic (75%) reactors. The major advantage of the MJB process is that no internal recirculation of biomass is required and the MLSS concentration along the process train is more uniform than the MUCT process, thus reducing total energy consumption. Figures 8.1 and 8.2 illustrate the process diagrams of the MUCT and MJB processes.

Even though the theoretical oxygen demand of the system can be estimated fairly accurately, aeration systems are typically designed to provide for complete COD removal and nitrification. In practice, the air supply can be turned down to provide for the actual oxygen requirement at a given time.

Care must be taken when selecting data for use in determining oxygen requirements. Inaccurate flow measurement, sample streams contaminated with recycle, non-flow-paced sampling, intermittent (one to three samples per week) sampling, and the laboratory test (5-day biochemical oxygen demand) used as the industry

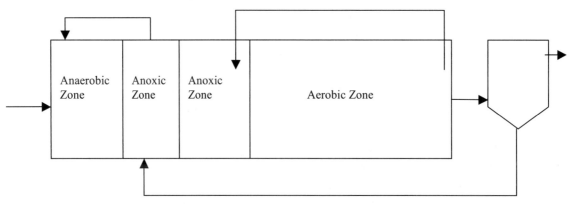

FIGURE 8.1   A diagram of the Modified University of Cape Town (MUCT) process.

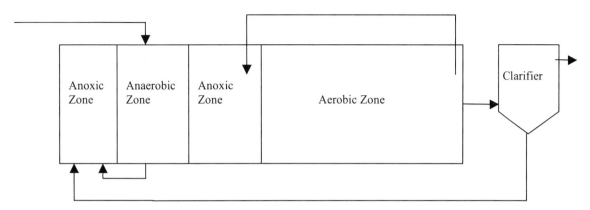

FIGURE 8.2 A diagram of the Modified Johannesburg (MJB) process.

standard may contribute to erratic data from which to try to develop meaningful relationships (Kennedy and Boe, 1985).

2.0 TYPES OF AERATION EQUIPMENT

Aeration exposes water or wastewater to oxygen so that it can be transferred into the fluid. A primary use for aeration (oxygen transfer) is to provide oxygen to microorganisms in wastewater treatment. Aeration may also be used to provide mixing in aerated grit chambers, increase dissolved oxygen in the treated water prior to discharge (post-aeration), or oxidize dissolved metals, remove entrained gases, or improve the dissolved oxygen content in drinking water treatment.

Aeration usually involves exposing water to air, but it can also involve exposure to high-purity oxygen in an enclosed tank. Increasing the gas-liquid interfacial area increases the oxygen transfer rate. A variety of devices have been developed and used to accomplish aeration. These have been discussed at length in references such as *Design of Municipal Wastewater Treatment Plants* (WEF and American Society of Civil Engineers, 2009) as well as *Wastewater Engineering, Treatment and Reuse* (Metcalf & Eddy, 2003). This section provides a brief overview of aeration devices.

2.1 Surface

2.1.1 Low-Speed Aerators

A low-speed aerator consists of a large-diameter impeller mounted on a vertical shaft that splashes the mixed liquor from an aeration basin into the air. Oxygen is transferred from the air into the water droplets as they come in contact. The aerator is usually mounted on a fixed concrete platform in the center of the aeration basin and driven through a gear box by an alternating current motor accessible at the platform level. Power draw increases significantly even with small increases in the impeller submergence depth. The oxygen transfer rate also increases with the impeller submergence depth. Therefore, to optimize oxygen transfer and power consumption, some facilities have been built with motor-operated weirs that reduce impeller submergence based on the dissolved oxygen concentration in the basin.

2.1.2   High-Speed Aerators

High-speed aerators, often associated with lagoons, are typically mounted on floats held in place by guy wires anchored at the edge of the lagoon. The power supply is brought to the aerator by a cable that can be submerged in the wastewater. The motor is directly coupled to the mixer impeller, both of which are on a vertical axis. The impeller splashes the wastewater into the atmosphere, resulting in oxygen transfer from air into the water droplets. Aeration is controlled through an operator-adjusted timer depending on daily or seasonal flow and load variation. The float rises and falls with the liquid depth in the lagoon; the submergence of the impeller is relatively constant. Operation of these units must be carefully monitored during cold weather, as ice can result in equipment damage.

2.1.3   Aspirating Aerators

Aspirating aerators draw air into the water in a low-pressure zone created by imparting high velocity to the water. The tube could be a part of the aerator, with an air inlet above the water surface. Alternatively, the device may use a submersible pump attached to a vertical air tube open to the atmosphere. Oxygen is transferred from the air contained in the bulk fluid to the fluid. These are low-cost mixing devices that can transfer oxygen rapidly, but at low-energy efficiency. Aspirating aerators not only have low aeration efficiency, they are limited in their oxygen transfer capacity (lb O_2/hr).

2.1.4   Brush Aerators

Brush aerators are most commonly associated with oxidation ditches. The aerator is mounted on a horizontal axis, which extends the width of the ditch. Both ends of the aerator are held in place by a bearing. One end extends and is connected through a belt or gear drive to an electric motor. The aerator consists of a shaft or drum with tines (finger-like strips of metal or other devices) that extend into the water. As the aerator revolves around the axis, the tines pick up small quantities of water and splash into the air. Oxygen is transferred from the air to the water droplets. At the same time, the tines impart a velocity to the water, moving it around the oxidation ditch. The water exposed to the atmosphere absorbs oxygen as the water returns to the moving body of fluid. As with low-speed aerators, the energy draw on the motor increases with small increases in liquid depth resulting from flow increases. This also results in an increase in oxygen transfer. Oxidation ditches have also been constructed with motor-operated, adjustable weirs to allow the operator to reduce aerator submergence.

2.1.5   Disk Aerators

Disk aerators are a variant of the brush aerators used in oxidation ditches. Instead of a drum with tines, the aerator consists of an axle with disks. As the rotor turns, the disks impart velocity to the water. Some of the water is splashed into the atmosphere and absorbs oxygen.

2.2   Submerged

2.2.1   Diffusers

Diffused aeration delivers compressed air through porous or perforated diffusers. Oxygen is transferred from the air bubbles to the bulk fluid. Diffusers are categorized as coarse bubble (typical bubble size >5 mm) or fine bubble (bubble size <5 mm). Both types are described in the following subsections.

2.2.1.1   Coarse Bubble

Coarse-bubble diffusers were commonly used for aeration from the 1950s through the 1970s. The energy crisis of the 1970s resulted in reevaluation, development, and application of fine-pore diffusers. Today, coarse-bubble diffusers are primarily used at small WWTPs, in anoxic selector zones, and in aerobic digesters. These diffusers are typically fabricated of heavy gauge polyvinyl chloride (PVC) or thin gauge stainless steel. Large holes are punched along the sides to give a wide airflow range.

2.2.1.2   Fine Pore

Fine-pore diffusers generate smaller air bubbles, increasing the interfacial area between air and water, resulting in higher oxygen transfer rates. The higher transfer rates reduce the need to compress air, thereby reducing the cost of operating aeration blowers.

Fine-pore diffusers are most commonly made of porous ceramic material supplied as discs or domes, or perforated membrane material. Ceramic diffusers, once supplied as a square plate, are now more commonly supplied as discs or domes. Membrane diffusers are available as circular inserts to a disc holder, horizontal tubes, or panels installed on the aeration basin floor. Disc diffusers are commonly mounted on PVC pipes and arranged in a grid pattern across the aeration basin floor. Tube diffusers are generally available in a variety of lengths and can be installed in a greater density than discs. Tube diffusers can usually be mounted closer to the basin floor, which can be useful in achieving desired OTE when retro-fitting diffusers in an existing shallow aeration basin. Panel diffusers are highly efficient because they can cover a large floor area. Panel diffusers are available in both wide and narrow (strip) configurations.

Fine-bubble diffusers generally have a relatively narrow operating range in which uniform airflow and bubble pattern across the surface can be achieved. As the airflow is reduced, the perforations at the periphery are subjected to greater closing forces due to the tendency of the disc to deform upward at the center, resulting in more air near the center. This can be compensated by making the membrane thicker in the center and thinner toward the periphery and/or by using a variable perfora-tion to promote more uniform air release. It is important to note that fine-pore dif-fusers made from flexible membranes can produce bubbles of varying diameter depending on the airflow. Thus a "fine-pore" membrane diffuser can produce "coarse bubbles" (>5 mm diameter) at high flow. In comparison, rigid-pore fine dif-fusers (ceramic) tend to produce "fine bubbles for a wider range of air flows."

A consideration with the installation of fine pore diffusers is an allowance for increased maintenance. Experience has shown that porous ceramic diffusers require ongoing maintenance and may need to be removed to provide adequate cleaning to restore transfer efficiency or reduce operating pressure. Membrane diffusers need maintenance and periodic replacement. As with ceramic diffusers, membrane dif-fusers are subject to fouling that deteriorates OTE; periodic flexing can address sev-eral types of fouling. The expected life for membrane diffusers is generally 5 to 10 years, depending on the conditions of operation.

2.2.2   Sparger Turbines

Sparged turbines are a combination of mechanical aerator and diffused aeration (U.S. EPA, 2000). The most visible part of this aerator is the motor and associated gear box to drive a vertical mixer, mounted at some depth below the water surface. Just below the mixer impeller is a sparger ring. Air is introduced through a blower and dis-charged through the sparge ring below each mixer. As air bubbles pass through the impeller, they are sheared into smaller bubbles and dispersed throughout the tank. Sufficient air is provided to meet oxygen requirements and the turbine is used for mixing. Sparged turbines have a limit as to the amount of air that can be sparged because excessive air floods the turbine blades, resulting in loss of pumping capacity and oxygen transfer.

2.2.3   Submerged Aerator Mixers

These are composed of a submerged motor with turbine impeller assembly, and an inlet line for air from a surface blower. These units are designed to operate at low speeds and have wide impellers to resist fouling. The rotating impeller shears the air into the water. During periods of reduced biological demand, the flow of incoming air can be adjusted while the mixing action remains on.

2.2.4   Static Tubes

Although static tube aerators are commonly associated with lagoons, they may also be used in basins. These are typically coarse-bubble systems. In these systems, air flows from an orifice in the bottom of an aeration header along the tank bottom. Above each orifice is a tube, about 0.30 m (12 in) in diameter and 0.75 m (30 in) in height. Inside each tube are static mixers (plates or tines) that break the air into smaller bubbles as it rises. The upward flow of air acts as an air lift to impart velocity to the bulk fluid. The system is simple and has no moving parts, but exhibits low OTE.

2.2.5   Jet Aerators

Jet aerators use water jets operated by pumps outside the aeration tank to circulate wastewater and mix the tank contents. Compressed or ambient air is introduced from a separate pipe attached to the water pipe and ejected through a nozzle assembly. As the air comes out, it diffuses into small bubbles, causing oxygen transfer from air to water.

2.2.6   U-Tube Aerators

U-tube aerators consist of a shaft that is 9 to 150 m (30 to 500 ft) deep. Air is added to MLSS as it enters the zone dedicated to the downward flowing liquid. The air and water travel to the bottom of the tube, where the air is forced into the liquid at great pressure. The mixture then flows upward through the return zone. These systems are efficient and do not have a large footprint, but are expensive to construct. The aerators are also particularly useful for high-strength wastewaters.

3.0   DESIGN CONSIDERATIONS

The type of energy conservation measure (ECM) used to conserve energy of the biological treatment process will be determined by the size of the plant, the strength of the wastewater, the local cost of electricity, and the type of process in use at the facility. Energy conservation measures are defined in Chapters 1 and 11 of this manual.

Lagoons, which are used by the smallest communities or industries, rely on the natural transfer of oxygen from the atmosphere to the bulk fluid. The relatively large surface area of most lagoons allows a sufficient transfer rate of oxygen for the demand to be met. As the load to a lagoon increases, additional cells may be added, or aerators may be added to increase the amount of oxygen that can be transferred into the water. As higher loads are experienced, special diffused aeration systems may be used. For the most part, lagoon systems use little energy. A timer may be installed on floating aerators or the blowers used to supply a diffused air system. Solar- or wind-powered water circulation devices can also be used to enhance oxygen transfer, although they are limited by the natural diffusion of oxygen from air at the water surface.

Oxidation ditches provide a reliable level of treatment and can be used to accomplish nitrification and/or BNR. Smaller installations (less than 50 hp) may consider the use of premium efficiency mixers, control of discharge weirs, and/or timers on the power supply to the rotors to control energy consumption. Larger installations may use dissolved oxygen measurement and feedback control to vary the output of a variable-frequency drive on the mixer motor. Control of the effluent weirs and premium efficiency motors may also be a consideration. At least one manufacturer provides a floating rotor that is not affected by the height of the effluent weir relative to the flow through the tank.

Plants that use biotowers must focus energy efficiency efforts on pumps used to lift the wastewater onto the tower. Likewise, trickling filter plants focus on the energy efficiency of pumps used to recirculate wastewater flow. Premium efficiency motors should be considered when pump motors are in need of replacement. Right sizing of pumps and proper trimming of pump impellers can be considered.

Plants that use high-purity oxygen may control the flow of oxygen to the aeration basins based on the level of carbon dioxide in the waste gas at the end of each aeration train. Alternatively, plants monitor the percent of oxygen in the waste gas and reduce (increase) the venting of gas based on the oxygen content compared to a setpoint value. Other ECMs include premium efficiency motors on aeration basin mixers, premium efficiency mixer design, and premium efficiency motors on oxygen production equipment.

Plants with mechanical aeration must rely on proactive maintenance of the gear boxes and premium efficiency motors. Observations indicate that plants using the concentration of dissolved oxygen to control the level of effluent weirs typically are not able to maintain the systems in operation and have to use frequent manual adjustment of effluent weirs based on operator experience.

Plants with diffused aeration will be the focus of the following discussion. There are three primary components to the systems: diffusers; aeration blowers; and instrumentation and control.

Industry is rapidly phasing out older, inefficient aeration systems and moving toward more efficient diffused air systems. The primary purpose of aeration equipment is to transfer the oxygen from the air/gas phase into the liquid to satisfy the process oxygen demands. In addition, aeration is used to provide sufficient mixing energy to keep MLSS in suspension. When designing these systems, the focus should be on the following items that influence power costs:

- Bubble (droplet) size;
- Bubble (droplet) density;
- Control over oxygen used by microorganisms;
- Optimum availability of oxygen to microorganisms;
- Optimum operating range of diffusers; and
- Optimum operating range of blower.

There is no single solution for every situation. Existing conditions at the site, competing interests among suppliers (blower/diffuser/control), and commercial bidding procedures often result in system selection based on price rather than energy efficiency. Lowest capital cost, however, may not necessarily be the best value considering life cycle cost.

Factors that must be considered for optimizing aeration system design to improve energy efficiency are presented in this section.

3.1 Oxygen Transfer Efficiency

Oxygen transfer efficiency under field conditions (OTE_f) is the mass of oxygen transferred to the aerated liquid from the mass of oxygen supplied. Oxygen transfer efficiency under field conditions depends on a number of factors including site elevation, temperature, dissolved oxygen saturation concentration, the type and installation of aeration equipment, characteristics of the wastewater, the treatment process, and the degree of treatment already received. Conditions may differ even between aeration tanks at a given WWTP. For example, OTE increases with increasing depth, increasing density of diffusers, decreasing diffuser flux rate, and increasing degree of treatment. The oxygen transfer rate under process conditions is determined by the following equation, as given in the *Design of Municipal Wastewater Treatment Plants* (WEF and American Society of Civil Engineers, 2009):

$$OTR_f = \alpha FSOTR\theta^{(T-20)}[(\beta\Omega\tau C^*_{\circ\,20}-C)/(C^*_{\circ\,20})] \qquad (8.7)$$

Where

OTR_f = oxygen transfer rate under process conditions, at an average dissolved oxygen concentration of C in the reactor and at temperature T (lb/hr);

α = average process water volumetric mass transfer coefficient KL_a/average clean water volumetric mass transfer coefficient KL_a (both with new diffusers);

$SOTR$ = standard oxygen transfer rate of new diffuser (lb/hr);

F = process water SOTR of a diffuser after a given time in service/SOTR of new diffuser in the same process water;

θ = empirical temperature correction factor equal to 1.024, unless the aeration system shows a different factor;

β = process water C^*_{st}/clean water C^*_{st};

Ω = pressure correction factor, for tanks less than 6.096m (20 ft) = barometric pressure under field conditions (kPa or psia)/Standard barometric pressure (101.3 kPa or 14.7 psia);

$C^*_{\circ\,20}$ = tabular value of dissolved oxygen surface saturation concentration at 20 °C, barometric pressure of 101.3 kPa (1 atmosphere), and 100% relative humidity (mg/L);

τ = a temperature correction factor for dissolved oxygen saturation (C^*_{st}/C^*_{s20});

C^*_{st} = tabular value of dissolved oxygen surface saturation concentration at actual process water temperature, a barometric pressure of 1 atm, and 100% relative humidity (mg/L); and

C^*_{s20} = tabular value of dissolved oxygen surface saturation concentration at 20 °C water temperature, a barometric pressure of 1 atm, and 100% relative humidity (mg/L).

To compare efficiencies of various aeration systems and equipment, standardized test procedures have been established (American Society of Civil Engineers, 2007). Shop tests are performed under standard conditions of clean deoxygenated water, with results reported under standard conditions of 20 °C, 1 atmospheric pressure, and 1000 mg/L total dissolved solids (TDS). Any deviations from these conditions are accounted for by applying appropriate correction factors.

Operating aeration equipment under standard test conditions yields the standard oxygen transfer efficiency (SOTE). The energy used per unit weight of oxygen delivered is the standard aeration efficiency (SAE). The actual oxygen transfer efficiency (AOTE) or OTE_f differs from SOTE because wastewater characteristics differ from those of clean water used in standard test conditions. The AOTE is influenced by water quality, particularly the presence of surfactants (surface active agents), such as detergents and dissolved and suspended solids. The properties of detergents that cause foaming or the formation of bubbles that resist collapsing also inhibit oxygen transfer, even in dilute solutions such as wastewater. Surfactants change the surface tension properties of the liquid phase, which, in turn, reduces oxygen diffusion.

3.1.1   Mixing

A minimum level of power per unit volume is required to ensure adequate mixing. The amount of power required depends on the tank geometry, the density and type of diffuser, and diffuser layout. For most aeration tanks, at least 3 to 15 W/m^3 (15 to 75 hp/mil gal) of tank volume will be required to prevent settling of MLSS. For fine-bubble aeration systems, the typical recommended airflow for design is 2.2 m^3/h per m^2 (0.12 scfm/ft^2) of the tank floor area, although, in practice, lower airflows have been successful and should be used if experience demonstrates adequate mixing is provided. During low aeration requirements (low flow, winter weather), mixers may be used to create velocities needed to overcome settling of MLSS. Generally, the mixer is positioned to direct flow across the surface of the diffuser. When mixing requirements far exceed process oxygen requirements, the operator may consider reducing the number of tanks in operation. Thus, when one or more tanks are taken out of service, the energy required to mix the remaining tanks is adequate to fulfill the process oxygen requirements.

3.1.2   Diffuser Flux Rate

The diffuser flux rate is the rate of airflow per unit surface area of the diffuser. With most diffusers, a minimum flux rate is needed to ensure a uniform distribution of air

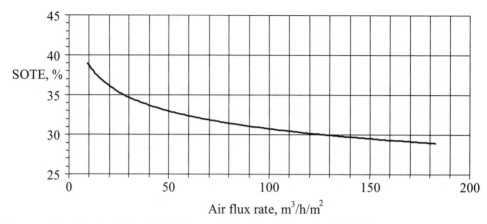

FIGURE 8.3 Typical Standard Oxygen Transfer Efficiency (SOTE) curve for a high-efficiency flexible membrane diffuser at 4.57 m diffuser submergence (courtesy of EDI).

throughout the perforated area of the diffuser. A high flux rate results in a decrease in efficiency as air bubble size increases, as shown in Figure 8.3. Fine-pore diffusers typically have the narrowest range of acceptable flux rates and provide the greatest oxygen transfer per kilowatt-hour (horsepower-hour) at their lowest acceptable airflow.

Oxygen transfer efficiency can be improved by reducing the flux rate. For a given airflow rate, increasing the diameter of the diffuser (if available) or the number of diffusers would result in an increase in the total diffuser surface area in an aeration tank, which would decrease the flux rate and improve OTE.

Oxygen transfer efficiency increases as the diffuser submergence depth increases. An air bubble released from the same diffuser at a greater depth spends more time in the water column before being discharged to the atmosphere. Another reason for OTE increasing is that the partial pressure of oxygen in the bubble increases with depth. Although OTE increases as depth increases, the required blower pressure for operation also increases; therefore, the net effect on SAE (kg O_2/kWh) remains relatively constant. The OTE of coarse-bubble diffusers typically increases slightly with increased airflow, although the OTE of coarse-bubble diffusers is less than that for fine-pore diffusers (U.S. EPA, 1989).

3.1.3 Alpha

The alpha factor is one of the most crucial design parameters in energy-efficient aeration. It is the ratio of the oxygen transfer coefficients in wastewater versus clean water, as shown in Eq. 8.7. The alpha factor is used primarily to determine the amount of air needed for a process. Alpha can be determined fairly accurately under process conditions by using techniques to capture the off-gas from operating facilities (Redmon et al., 1983). The application of this technique was demonstrated for investigating OTE of a small-scale system (Mahendraker, Mavinic, and Rabinowitz, 2005). In plug-flow aeration tanks, alpha increases from the inlet to the outlet as the offending contaminants are biodegraded (U.S. EPA, 1989). The alpha factor variation is most pronounced in fine-pore systems; it may be 0.4 or lower at the inlet and increase to 0.8 or higher at the outlet of a plug-flow reactor, varying as the organic matter is removed from the wastewater. Historically, it was argued that the average alpha could be increased by increasing the MLSS and operating at a lower F/M ratio to more rapidly remove the surfactants. Doing so, however, increases oxygen demand by higher endogenous respiration. In addition, plants that do not have a nitrification or nitrogen removal requirement will encounter higher oxygen demand from nitrification. The operator must also be concerned with development or aggravation of *Nocardia* or other types of filamentous bacteria that cause foam or sludge bulking. Reducing the loading rate may result in aeration being governed by mixing requirements, operating at higher dissolved oxygen concentrations, and reducing the driving force of oxygen to the water. An alternative strategy is described in Section 4.3.

3.1.4 Beta

The beta factor is the ratio of the saturation concentration of oxygen in wastewater to that in clean water. This is also important in determining the actual amount of air needed for a process. This factor can vary from a low of 0.9 to a high of 0.99. Salinity and alkalinity are the main factors that affect beta. Beta can be estimated from TDS as follows (Mueller et al., 2002):

$$\beta = 1 - 5.7 \times 10^{-6} \times \text{TDS} \tag{8.8}$$

The operator can do little to increase the beta factor cost effectively. Incorrectly identifying a low beta, however, can lead to undersizing of the blowers.

3.1.5 System Costs

Mueller et al. (2002) compiled the capital and operating costs of aeration systems in activated sludge plants, which clearly indicate the impact of aeration systems on overall costs. These are presented in Table 8.1.

TABLE 8.1 Aeration system costs (Mueller et al., 2002)

Plant Name	Location	Design Flow, m³/d (MGD)	Type of Aeration System	Capital Costs		Yearly Operating Costs	
				Total Plant, $M (Year)	% Due to Aeration	Total Plant, $M (Year)	% Due to Aeration
Coney Island	Brooklyn, NY	378 541 (100)	Diffused, Fine Bubble	650 (1990)	20	4.43 (1998)	20.1–25.5*
						4.05 (1999)	20.3–25.2*
North River	Manhattan, NY	643 520 (170)	Diffused, Fine Bubble	968 (1986)	5.57	7.12 (1998)	15.7
						7.43 (1999)	16.8
Red Hook	Brooklyn, NY	227 125 (60)	Diffused	232 (1988)	16.8	2.49 (1998)	25
						2.29 (1999)	24
Owls Head	Brooklyn, NY	454 249 (120)	Diffused	380 (1995)	27	7.15 (2000)	17
West Point	Seattle, WA	503 460 (133)	High Purity Oxygen, Surface, 4-Stage	229 (1995)	19.3		
Middlesex County Utilities Authority	Sayreville, NJ	556 456 (147)	HP, Turbine, Surface	95.5 (1974)	19.3	16.4 (1997)	19.5 (before upgrade)
				+8.9 (1995)	100	15.2 (1999)	13 (after upgrade)
Darmstadt Central	Germany	37 854 (10)	Diffused, Fine Tubes with Propellers, Racetrack	95 (1995)	15	3.4 (1997)	11.4

* Including air scrubbers

4.0 OPERATIONAL CONSIDERATIONS

4.1 Waste Loading Distribution

Among activated sludge process modifications, some are defined by the distribution pattern of waste loading to the aeration tank. For example, plug-flow tanks receive inflow at the upstream end, whereas step-feed tanks distribute the wastewater feed along the length of the tank. Complete-mix activated sludge produces a uniform waste loading throughout each aeration tank. The degree of treatment increases uniformly along the flow path in a plug-flow aeration tank and in stages in a step-feed tank, but it remains uniform throughout complete-mix aeration. As indicated previously, OTE improves with the degree of treatment; that is, OTE tends to approach SOTE.

4.2 Step-Feed/Complete-Mix Modes

Operating in step-feed or complete-mix modes may result in a more uniform overall alpha by distributing surfactants throughout the aeration tank and allowing for their dilution and more rapid removal by adsorption (U.S. EPA, 1989). Another possible benefit for a fine-pore system is the reduction of diffuser fouling, which is known to occur at the inlet-area diffusers in plug-flow systems.

4.3 Mixed Liquor Dissolved Oxygen

The most common cause of aeration system inefficiency is greater than necessary dissolved oxygen concentration in the mixed liquor, which is generated at the expense of high-energy use. The closer the dissolved oxygen is to saturation, the greater the resistance for dissolution of oxygen will be and, hence, the lower the OTE (see Eq. 8.7).

Oxygen transfer from air to water is a mass transfer reaction, directly proportional to the driving force, which is the difference between the dissolved oxygen saturation concentration and the dissolved oxygen in bulk liquid. As oxygen is transferred from air to water, the dissolved oxygen concentration in bulk liquid increases and the driving force consequently decreases. Thus, to optimize oxygen transfer efficiency, the operator must operate the aeration tank at the minimum dissolved oxygen concentration to achieve process objectives. At the same time, solids must also be maintained in suspension. The impact of potential savings achieved when operating at a given dissolved oxygen concentration compared to the set-point is presented in Figure 8.4. The equation describing the oxygen transfer under process conditions is given as:

$$dC/dt = K_L a(Cs-C) - r_M \qquad (8.9)$$

Where

  Cs = dissolved oxygen concentration in equilibrium with gas as given by Henry's law (mg/L);

  C = dissolved oxygen concentration the liquid or reactor (mg/L); and

  R_M = rate of oxygen use by the microorganisms.

The saturation values for diffused-air aeration systems is a function of liquid temperature, atmospheric pressure, diffuser submergence (also pressure related), and

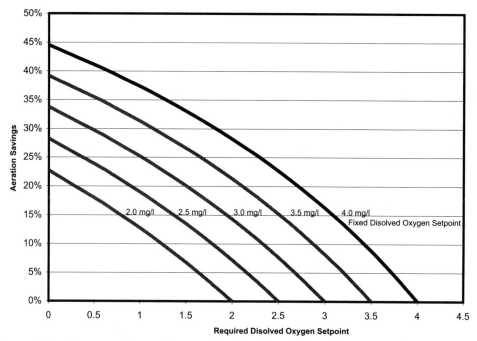

FIGURE 8.4   Energy savings with operating level vs. set-point (courtesy of Mathew Gray).

dissolved salts (TDS). Saturation values for surface aerators are affected by the same parameters except submergence.

When manual control is practiced, operators will tend to overcompensate to "keep ahead of the game" should an increase in load occur before the next mixed liquor dissolved oxygen reading. Installation of online dissolved oxygen meters with instrumentation can provide control of the aeration blowers so that the supply of oxygen increases with increasing demand by the microorganisms, avoiding the occurrence of excessive mixed liquor dissolved oxygen. Evolution in the technology for online metering and controls has resulted in minimum maintenance to achieve reliable results. In medium-to-large WWTPs, the payback period for installing automated blower control from mixed liquor dissolved oxygen probes is generally within a few years.

The following ECM is recommended:

- Installation and use of dissolved oxygen probes in aeration tanks with blower feedback control systems can result in substantial energy and cost savings with quick payback.

Figure 8.5 illustrates the potential for reducing the energy required for aeration by online monitoring and control of dissolved oxygen concentration in the aeration basin. The solid lines in Figure 8.4 show profiles of dissolved oxygen during a 24-hour test period in two parallel, completely mixed aeration trains. Lines labeled *manual* and *automatic* represent oxygen concentrations in aeration tanks with manual and automatic control of dissolved oxygen, respectively. With the exception of the evening hours, the line representing manual control is higher than the line representing automatic dissolved oxygen control. This example illustrates the potential that automatic controls have to match the concentration of dissolved oxygen to the target level. The estimated energy savings with automatic controls averaged 33%. Actual energy savings realized with automatic controls were higher, at 38%.

The location of the dissolved oxygen probe will affect the ability of the system to identify and respond to changes in demand for oxygen in a timely manner. If located at the end of a plug-flow aeration basin, the dissolved oxygen sag caused by a peak loading would not be observed until well after the peak has entered the tank.

In plants with multiple-pass aeration basins, one dissolved oxygen meter would be needed in each pass, as shown in Figure 8.6. Hand-held dissolved oxygen probes can be used to ensure that all the basins are operating in a similar manner.

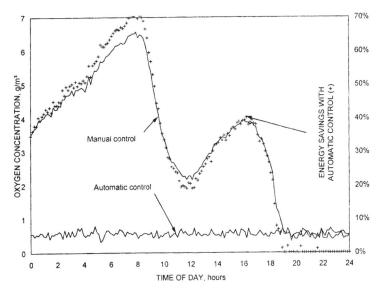

FIGURE 8.5   Comparative dissolved oxygen profiles for automated control versus manual operation (U.S. EPA, 1989).

There are two ways to control single-pass aeration basins constructed in parallel. One would be similar to the system shown in Figure 8.6, where one dissolved oxygen meter would be provided per basin and airflow controlled to each basin. Manual flow control valves would be provided for each drop to a grid of diffusers.

The second approach is to install dissolved oxygen meters in the inlet, middle, and outlet aeration grids, as shown in Figure 8.7. In this approach, the automatic control system can better adjust to spatial variations in oxygen demand that occur during the day. Again, hand-held probes are used to monitor the parallel basins so manual adjustments can be made if necessary.

Using two or more probes per tank can lead to greater control, but the costs associated with the probes, control system, and maintenance reach diminishing returns. Multiple probes may only prove of value in larger systems. The financial return of using multiple dissolved oxygen probes for distributing oxygen to locations that

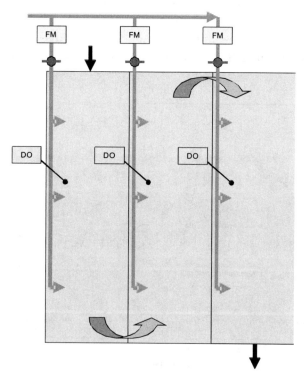

FIGURE 8.6   Dissolved oxygen probe location—Scheme 1 (courtesy of Jim Marx).

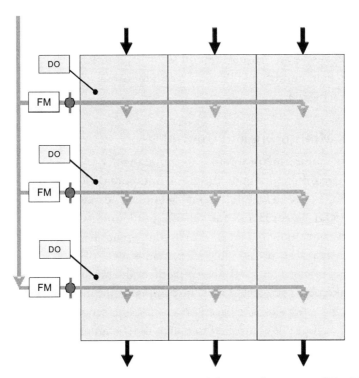

FIGURE 8.7 Dissolved oxygen probe location—Scheme 2 *(courtesy of Jim Marx).*

need it is also dependent on local power costs. In some parts of the United States, power costs are low and, therefore, the cost benefit of doing so may be limited.

Most WWTPs have characteristic daily, weekly, and seasonal loadings. With the appropriate analysis and study and some round-the-clock sampling and analysis, the wastewater can be characterized. Using this characterization study, an operator can predict increases or decreases in COD load and can compensate for this change by making the appropriate blower adjustment in anticipation of the changes rather than after the fact. This approach can dramatically improve both treatment performance and aeration efficiency. It will help improve energy efficiency for either manually controlled or automatically controlled systems.

The following ECMs are recommended:

- Characterize the waste stream to the aeration system to anticipate loading changes by conducting round-the-clock sampling studies.

- Technologies have been developed that use off-gas analysis to determine the actual use of the oxygen supplied to a given basin area. These systems are more expensive and require careful setup, calibration, monitoring, and adjustment to field conditions. Paybacks of 6 months to 2 years in energy savings may be achieved.

4.4 Process Monitoring and Control

A good process monitoring program is essential to energy efficiency of aeration systems. Ongoing maintenance is required for fine-pore diffusers to maintain OTE as well as operating pressure within reasonable limits. The manufacturer of the installed equipment should be contacted for specific guidelines. Constant process testing, operating plan re-evaluation, and careful implementation of operating strategies are needed. This is a cyclic process. Table 8.2 provides ideas of observations that may be made, the possible cause for conditions outside of the norm, and examples of actions that may be taken. It is suggested that the original equipment manufacturer be contacted for specific actions to take based on specific equipment installed in the plant.

Ideally, air delivery is controlled to match the instantaneous oxygen demand at representative points in the aeration tank. This requires monitoring dissolved oxygen

TABLE 8.2   Examples of conditions/causes/actions

Observation	Possible Cause	Actions
Coarse bubbling in a fine-pore aeration grid	Biological fouling or high surfactant concentrations in wastewater	Drain basin to inspect diffusers. Investigate local industries for discharge of cleaning agents.
Increasing blower energy use or rising air pressure	Onset of fouling	If equipment is available, check the dynamic wet pressure across the diffuser.
Changes in tank dissolved oxygen profiles	The onset of biological changes that need to be addressed or improperly set distribution valves	Check flow and organic loading to the aeration basin and check key operating parameters. Check operation of automatic valve operators and/or manual valve operators. If membrane diffusers, try flexing diffusers to free of fouling.
Aeration tank surface patterns can reflect the distribution of air and the size of air bubbles	Membrane or piping failure	Monitor air rates, dissolved oxygen values, and organic loading to identify changes in operating parameters. If membrane of pipe failure is indicated, plan and conduct repairs.

concentrations at selected points in the mixed liquor. Operation involves continual adjustment of air delivery so that oxygen concentrations remain near the target level.

The target oxygen concentration is the level consistent with satisfactory operation of the activated sludge process. It is determined by test and by experience for a particular WWTP to sustain the oxygen requirements of the activated sludge flora. For activated sludge processes, the target level is typically 1.5 to 2 mg/L. Automated process controls can typically match air delivery to oxygen demand more closely than manual controls.

Dynamic dissolved oxygen setpoints that vary with diurnal loading (higher dissolved oxygen during high loading, lower dissolved oxygen during low loading) can produce aeration savings between 10 and 20% compared to fixed dissolved oxygen setpoints. The dynamic dissolved oxygen setpoints can be controlled by using online ammonia analyzers in a feedback or feed-forward control loop (feed forward is preferred over feedback as feedback may miss the peak loadings) (International Water Association, 2006).

4.5   Dissolved Oxygen Management in Membrane Bioreactor Systems

In addition to biological process aeration, membrane bioreactor (MBR) systems require air for scouring of the membranes. The objective is to control membrane fouling by air scouring to meet the flow capacity of the system at all times. The membrane technology supplier invariably defines the frequency and airflow required for membrane scouring. The design of the membrane aeration system is also specific to the membrane supplier, which is typically supplied along with the membrane units. In an analysis by DeCarolis et al. (2008), the energy demand for membrane scouring was found to be 38% of the total energy demand.

In practice, the dissolved oxygen in the membrane tank and RAS can be as high as 5.0 mg/L. By designing a suitable RAS feeding point within the process train, it is feasible to use the residual dissolved oxygen to optimize the total aeration requirements of the process. For example, when a Modified Ludzack-Ettinger (MLE) process is applied to treat municipal wastewater with the goal of achieving effluent total nitrogen of less than 10 mg/L, the RAS can be fed to the aerobic reactor to use the residual dissolved oxygen in the mixed liquor. Nevertheless, an additional internal biomass recirculation for nitrates, between the aerobic reactor and the anoxic reactor, would be required for denitrification. If low effluent total nitrogen (e.g., 5 mg/L) with or without external carbon is required through the

MLE process, then a de-oxygenation and denitrification zone can be added within the RAS loop for use of the residual dissolved oxygen by endogenous respiration and partial denitrification. Subsequently, RAS without dissolved oxygen is fed to the anoxic reactor to maximize denitrification and accomplish low-effluent total nitrogen concentration.

Effluent permits often specify dissolved oxygen requirements (e.g., 5 to 6 mg/L) and, because MBR effluent (permeate) has similar levels of residual dissolved oxygen, the energy for post-aeration is minimized. Furthermore, because the permeate BOD and ammonia-nitrogen concentrations are much lower than the effluent from a conventional activated sludge process, the oxygen content is not used and the overall impact on the receiving water is positive due to high residual effluent dissolved oxygen from an MBR system.

4.6   Fouling of Porous Diffusers

Fouling of porous diffusers can originate from the air side, water side, or both, and can be manifested as pore fouling or surface fouling. Pore fouling constricts pores. Surface fouling, conversely, produces the effect of enlarged effective pore sizes. As surface fouling increases, bubble size increases and coarse bubbling may ensue. Wastewater foulants on porous diffusers include

- Solids and salts in mixed liquor that may penetrate diffusers;
- Biological growths (bio-fouling);
- Mineral deposits, or fine sand on the outer surfaces of diffusers; and
- Solids from improperly filtered air or from flaking of internal air line surfaces on the underside.

Bio-fouling can also occur when diffusers are operated below their recommended minimum flux rate. Insufficient airflow does not produce enough activity on the membrane to inhibit biological growth on the membrane material itself. Some membrane materials have incorporated biocides within the compound matrix, whereas others have used chemical compounds that resist attached-biofilm growth.

Once bio-growth is established, it can continue to the point where the pores are almost completely plugged. In theory, this should not happen because accurate tracking of power consumption, blower motor amperage, or discharge pressure will provide warning that the diffusers are fouling in some manner.

Air-side foulants include particulates from improperly filtered air and other material originating after filtration such as oil from improperly maintained blowers or corrosion from the inside of air mains. Improved air filtration and the use of corrosion-resistant materials of construction have made air-side fouling of porous diffusers uncommon. However, some post-filtration foulants to consider include

- Diffuser system degradation products such as rust or flaked paint,

- Mixed liquor solids leaked into the air pipes, and

- Construction debris remaining from before the initial startup of the air delivery system or entering on subsequent reopening.

Fouling may reduce field OTE (increasing the amount of air needed from the blowers), increase the backpressure on the aeration blowers (increasing the power required to deliver the same amount of air), or both.

Evaluating diffusers for fouling is particularly critical after an operational upset. For example, a loss of air to the diffusers (especially ceramic diffusers) could result in intrusion of mixed liquor solids, causing the diffuser pores to become fouled. Loss of air to the diffusers may necessitate cleaning all diffusers. Cleaning diffusers after a power failure is seldom a trivial task. Reliable power and a standby supply can be cost-effective precautions against costly power failures. Many newer aeration devices include check valves close to the diffuser and small air plenums that limit the amount of mixed liquor that can penetrate the diffusers on loss of pressure.

4.7   Diffuser Cleaning

4.7.1   Air-Side Fouling

After construction or any opening of the air distribution system, and before installation of any porous media, it is essential to conduct a thorough purging or cleaning of the system. Air filters must be installed and maintained for the aeration system to work properly and to prevent air-side fouling. Poorly maintained viscous impingement air filters will create excess suction vacuum and possible duct leakage or increased power consumption and compressor maintenance. Once diffusers are fouled on the air side, there is no economical method for cleaning them; they must be replaced.

4.7.2   Liquid-Side Fouling

Diffuser cleaning is largely or entirely directed toward the water face of the diffusers. For in situ cleaning, the diffusers are cleaned in the tank. In situ methods are available for either full or drained tanks. In situations involving dewatering aeration tanks, personnel safety is paramount. Provisions must also be made to reduce the risk of damage to the diffuser system. Polyvinyl chloride pipe work and components are susceptible to damage by exposure to direct sunlight. Unforeseen and excessive thermal expansion and contraction while the tank is dewatered can also damage PVC components. Emptying tanks during temperature extremes should be avoided. Ice rafts may damage equipment, and falling influent may damage pipes or diffusers.

In situ methods include physical, chemical, and biological procedures. The diffusers may be physically washed with water, air, or steam, which dislodges loose external biological growths and deposits. Chemical procedures may include addition of gaseous compounds such as hydrogen chloride, chlorine, or gaseous biocides to the air side or may include liquid acid or detergent cleaning of diffuser surfaces. Hydrogen chloride 14% solution sprayed on diffusers accompanied by hosing before and after removes both organic and inorganic foulants. Some of these methods are patented, expensive, and can pose serious safety concerns. It is best to consult with the diffuser supplier to determine the best cleaning method as every material has properties that may be different from another of the same generic name and could respond differently to various treatment regimes.

For ex situ cleaning, the diffusers are removed from the aeration tank. Ex situ methods include refiring of ceramic diffusers, high-pressure water jetting, and washing ceramic diffusers with silicate-phosphorus, alkali, acid, or detergent. Refiring ceramic diffusers in a kiln destroys organic foulants, but may incorporate foulant residues in the porous matrix and leave behind ash that may plug or reduce flux capacity. Ex situ methods are unpredictable in their effectiveness and are expensive. Therefore, they tend to be used after in situ methods have failed to produce the desired results.

Some manufacturers of perforated membrane diffusers suggest weekly or monthly air bumping, also called *flexing*. First, the airflow is stopped to collapse the diffusers onto their frames. Next, the flow is increased to two or three times what is normal, but within the maximum rating of the diffuser. Finally, the airflow is restored to normal. In systems equipped with computerized controls, this function may be programmed. In manually operated systems, grids must be tuned down to minimal operation in order to redistribute the air to other grids to then redistribute the airflow to the appropriate area. Nevertheless, experience with such procedures has not always been successful. Brush scrubbing preceded and followed by hosing has been used to clean membrane diffusers.

Porous stone (ceramic) diffusers have been cleaned in situ by means of acid gas injection. This involves the injection of hydrogen chloride or formic acid gas into the air delivered to diffusers. The airflow is held near the maximum flow rating of the diffusers for uniform cleaning of their pores. Safety training and precautions are essential. Gas injection typically continues until the pressure differential across the diffusers has stabilized, and the pressure stabilizes after approximately 30 minutes of acid gas flow. The hydrogen chloride cleaning process is patented. It produces a hydrochloric acid solution of approximately 28% concentration in diffuser pores. The liquid acid dissolves some inorganic salts deposited in the pores and may help remove bio-films. However, the acid does not remove dust from the air side or silicic deposits from the wetted side.

Rigid, porous, plastic diffusers have been cleaned using many of the same options for ceramic diffusers, with the exception of refiring, flaming, and sand-blasting.

The choice of cleaning method is often determined by trial and error. Simple procedures such as hose washing are often tried first, with more difficult procedures following if the previous cleaning was unsuccessful or short-lived. When cleaning is short-lived, operators must resort to removal of diffusers for testing of various cleaning agents or physical procedures in the laboratory.

5.0 DIFFUSED-AERATION CASE HISTORIES

The following simplified summaries are restricted largely to the diffused-air system alone (Chann, 2008). Generally, these case histories document improved OTE with fine-pore aeration. In many instances, a cost advantage of fine-pore aeration is evident. In several instances, preliminary simple payback periods are computed. The payback period is the initial cost associated with fine-pore aeration divided by the annual savings obtained.

5.1 Batesville, Arkansas

This industrial facility in Batesville, Arkansas, operates an activated sludge process in an earthen reactor. The plant is designed to treat up to 9070 kg/d (20 000 lb/d) of BOD. Static tube diffusers were replaced with fine-bubble, flexible membrane tube diffusers using a floating lateral system. The static tube system required the operation of three 373-kW (500-hp) and one 522-kW (700-hp) blower. After the conversion, one 522-kW (700-hp) unit and one 373-kW (500-hp) unit were required on an intermittent basis (approximately 746 kW [1000 hp] on average). This resulted in a savings of 5.63×10^6 kWh per year. Based on a capital cost of approximately $1 million, the estimated payback term for the project was less than 3 years.

5.2   Beloit, Wisconsin

The City of Beloit, Wisconsin, operates an activated sludge plant that uses ceramic disc diffusers in the aeration basins. Plant loadings are such that only two of four aeration basins are operated for treatment. The remaining basins are operated in an idle condition. Ceramic diffusers have no check-valve feature and, to prevent intrusion of water/mixed liquor into the piping system, a minimum airflow of 0.85 sm³/h (0.5 scfm) per diffuser is required. There are more than 2600 diffusers in each basin, requiring a minimum airflow of 4420 sm³/h (2600 scfm) to maintain the diffusers in the two idle basins. The city converted the diffusers in the idle basins with flexible membranes. Membrane diffusers have backflow prevention capabilities and do not require a minimum airflow to maintain functional performance. The city was able to reduce the required airflow by approximately 4420 sm3/h (2600 scfm). This resulted in a yearly energy savings of 6.53×10^5 kWh per year.

5.3   Palmyra, Wisconsin

The Village of Palmyra, Wisconsin, operates a three-cell, aerated lagoon facility. The facility has a design capacity of 871 m³/d (0.23 mgd), and treats both domestic and industrial wastewater flows. The current flow to the plant averages 644 m³/d (0.17 mgd). Static tube diffusers were replaced with fine-bubble, flexible membrane tube diffusers using a floating lateral system. The energy use for the entire facility decreased from 47 000 kWh per month to 22 000 kWh per month after the conversion. The energy reduction was the direct result of the aeration system conversion as no other process or equipment changes were associated with the project.

6.0   MECHANICAL AERATION CONTROL

Mechanical aerators dissolve oxygen by thrashing the water surface to increase the surface area of water exposed to air. Means of adjusting oxygenation capacity to match the oxygen demand include submergence adjustment, speed adjustment, and on-off operation.

6.1 Submergence Adjustment

The relative immersion of surface mechanical aerator impellers affects the oxygen-transfer capacity of the aerator. Adjusting the immersion can help match aeration capacity to oxygen demand. Figure 8.8 presents the results of Russian tests of the energy efficiency impacts of adjusting submergence (WPCF, 1988). The rate of change observed at transition points presented by these results indicates the sensitive hydraulics of the mechanical aerators tested. Operation at points significantly less than the optimum condition can impair energy efficiency. Data from the Russian tests indicate that energy efficiency deteriorates as the submergence is moved from its optimum location.

Given appropriate controls and equipment, it is desirable to adjust aerator immersion to achieve energy efficiency by matching oxygenation capacity with oxygen demand, as is done by throttling blowers.

For fixed aerators, provision for controlling the liquid level is achieved by adjusting the height of the outlet weir on the tank. Some systems have attempted to do this with motor-operated weir positioners. For floating aerators, immersion is adjusted by adding or removing weights, which is not practical for achieving hourly or daily variations in demand for oxygen.

6.2 Speed Adjustment

Aerator power and oxygenation capacity are a function of impeller speed. Frequently, fixed mechanical aerators are equipped with two-speed motors, which provide the operator with oxygenation flexibility. Another method for achieving speed adjustment is by installing an adjustable speed drive on the drive motor in the aeration basin, with speed adjusted based on a feedback loop from a measurement such as dissolved oxygen.

Because available aerator characteristics suggest that energy efficiency may be relatively constant over a range of operating speeds, speed variation may have an energy efficiency advantage over submergence adjustment. The feasibility of speed variation versus submergence adjustment to control oxygenation is largely determined in the initial design of WWTPs.

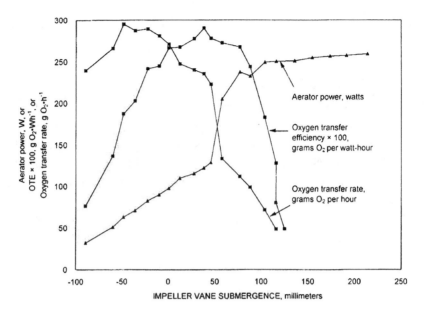

FIGURE 8.8   Mechanical aerator power, oxygenation efficiency, and oxygenation rate versus impeller submergence.

6.3   On-Time Adjustment

On-off switching of mechanical aerators may be the most practical way to adjust oxygenation capacity to oxygen demand, but it has pitfalls. Switching must not be so frequent that it would result in overload to the electrical equipment or controls. Nor should the aerator remain off for too long. Switching of multiple aerators should be desynchronized to avoid peak demand by having all mixers on at the same time.

Excessive off time may jeopardize the activated sludge process. A level of dissolved oxygen in the mixed liquor significantly below the target level may stress the microorganisms, or MLSS may settle. In general, at least 35 W/m^3 (75 hp/mil gal) of power must be provided to keep mixed liquor solids in suspension. Because of difficulty with even spacing of aerators in an aeration tank, a greater watts-to-cubic-meters (horsepower-to-million-gallons) ratio may be required to ensure adequate mixing in all reaches of the tank.

Each of these options requires proper design and operation. Process control considerations include the stability of the entire aeration system. Loading fluctuations, oxygen demand estimates, and the oxygenation capacity of each aerator all must be taken into account.

6.4 Mechanical Aerator Maintenance and Troubleshooting

Surface or structural damage to the impellers or shrouds will impair the original energy efficiency. Prompt identification and repair of damage and removal of entrapped material such as rope will help maintain energy efficiency.

Because some mechanical aerators are inseparable from their motors, scheduled lubrication and preventive maintenance of motors are essential. Regular maintenance according to manufacturer's recommendations is critical to energy efficiency and reliable operation. Personnel should note noise, vibration, and temperature routinely and conduct periodic electrical and mechanical checks.

Motors, especially those on floating aerators, can become coated with aerosols of the mixed liquor that dry out and form a surface coating. If the coating becomes too thick, it will act as an insulator on the motor's cooling fins and will lead to motor burnout if not removed routinely. Routine hosing of the motors should be a preventive maintenance activity and, as such, included in the preventive maintenance program. Icing is another concern in some colder areas of the country. To prevent straining supports or overloading motors, ice should not be allowed to accumulate.

7.0 REFERENCES

American Society of Civil Engineers (2007) *Measurement of Oxygen Transfer in Clean Water;* ASCE Standard No. ASCE/EWRI 2-06; American Society of Civil Engineers: New York.

Chann, R. C. (2008) Personal communication with Randall C. Chann of Environmental Dynamics, Inc., Columbia, Missouri (www.wastewater.com).

DeCarolis, J.; Adham, S.; Pearce, W. R.; Hirani, Z.; Lacy, S.; Stephenson, R. (2008) The Bottom Line - Experts Evaluate the Costs of Municipal Membrane Bioreactors. *Water Environ. Technol.,* **20** (1), 55.

Gujer, W.; Jenkins, D. (1974) A Nitrification Model for Contact Stabilization Activated Sludge Process. *Water Res.,* **9** (5), 5.

Hoover, S. R.; Porges, N. (1952) Assimilation of Dairy Wastes by Activated Sludge. II. The Equation of Synthesis and Rate of Oxygen Utilization. *Sew. Ind. Wastes,* **24**, 306.

International Water Association (2006) *Instrumentation, Control and Automation in Wastewater Systems;* International Water Association: London, United Kingdom.

Kennedy, T. J.; Boe, O. K. (1985) Efficient Aeration Operating Practices. Paper presented at the 58th Annual Conference of the Water Pollution Control Federation, October 6-10. Kansas City, Missouri.

Mahendraker, V.; Mavinic, D.S.; Rabinowitz, B.; Hall, K. J. (2005) The Impact of Influent Nutrient Ratios and Biochemical Reactions on Oxygen Transfer in an EBPR Process—A Theoretical Explanation. *Biotechnol. Bioeng.*, **91**(1), 22–42

Mahendraker, V.; Mavinic, D.S.; Rabinowitz, B. (2005) Comparison of Oxygen Transfer Parameters from Four Testing Methods in Three Activated Sludge Processes. *Water Qual. Res. J. Can.*, **40** (2), 164–176.

Metcalf & Eddy (2003) *Wastewater Engineering, Treatment and Reuse,* 4th ed.; McGraw-Hill: New York.

Mueller, J. A.; Boyle, W. C.; Popel, H. J. (2002) Aeration: *Principles and Practices;* CRC Press: Boca Raton, Florida.

Redmon, D. T.; Boyle, W. C.; Ewing, L. (1983) Oxygen Transfer Efficiency Measurements in Mixed Liquor Using Off-Gas Techniques. *J. Water Pollut. Control Fed.*, **55** (11).

Rosso, D.; Larson, L. E.; Stenstrom, M. K. (2008) Aeration of Large-Scale Municipal Wastewater Treatment Systems: State-of-the Art. *Water Sci. Technol.*, **57**, 973.

U.S. Environmental Protection Agency (2000) *Decentralized Systems-Technology Fact Sheet, Aerobic Treatment;* EPA 832-F-00-031; Office of Water: Washington, D.C.

U.S. Environmental Protection Agency (1989) *Fine Pore Aeration Systems;* EPA/625/l-89/023; Washington, D.C.

Water Pollution Control Federation (1988) *Aeration;* Manual of Practice No. FD-13; Water Pollution Control Federation: Alexandria, Virginia; Manuals and Reports on Engineering Practice No. 68; American Society of Civil Engineers: New York.

Water Environment Federation; American Society of Civil Engineers (2009) *Design of Municipal Wastewater Treatment Plants,* 5th ed.; Manual of Practice No. 8; ASCE Manuals and Reports on Engineering Practice No. 76; Water Environment Federation: Alexandria, Virginia.

8.0 SUGGESTED READINGS

Eckenfelder, W. W., Jr.; O'Connor, D. J. (1961) *Biological Waste Treatment;* Pergamon Press: Elmsford, New York.

Houck, D. H.; Boone, A. G. (1981) *Survey and Evaluation of Fine Bubble Dome Diffuser Aeration Equipment;* EPA-600/2-81-222; U.S. Environmental Protection Agency, Office of Research and Development: Cincinnati, Ohio.

Stenstrom, M. K. (2001) Aeration Systems, 20 Years' Experience; Civil and Environmental Engineering Department, University of California, Los Angeles. http://www.seas.ucla.edu/stenstro (accessed March 2009).

Chapter 9

Blowers

Proper consideration of air requirements and blower energy consumption is an essential part of both plant design and energy cost reduction in existing wastewater treatment facilities. The aeration blowers typically consume 50% of the total energy in diffused aeration activated sludge secondary treatment plants (see Chapters 7 and 8).

227

Blowers are also used in many other processes, including grit removal, channel aeration, aerobic digestion, and post-aeration. In water treatment facilities, blowers are most commonly used for filter backwash.

As important as blowers are to energy consumption, they are even more important to maintaining proper process performance. An adequate supply of air to the treatment process is essential to meet permit limits. Process compliance considerations should, in all cases, override energy conservation.

Blowers may be used for compressing gases other than air, such as methane for mixing anaerobic digesters. Many of the same types of equipment and application considerations apply for any gas being compressed. The blower manufacturer should be consulted for performance evaluations when compressing gases other than air.

1.0   APPLICATION CONSIDERATIONS

There are many similarities between pumps (see Chapter 4) and blowers. Positive-displacement and centrifugal designs are available for both. The energy consumption of both pumps and blowers is a function of flowrate, discharge pressure, and equipment efficiency. A variety of control methods, such as variable speed and throttling, can be used to control either pumps or blowers. A system analysis including process characteristics along with pump or blower characteristics is essential to performance optimization. However, the compressibility of air makes blower application and control more complex than pump application.

Detailed analysis of blower operation can become extremely cumbersome. A blower manufacturer's design and selection procedures include rigorous calculations and consideration of details that may not be significant in routine energy evaluations. The simplified formulas presented herein are sufficiently accurate for comparisons of various systems as well as to verify the cost-effectiveness of conservation measures. The blower manufacturer should be consulted for final confirmation of all designs.

1.1   Effects of Compressibility

Air is a mixture of gases. Dry air is predominantly nitrogen (75.5% by weight) and oxygen (23.1% by weight), with the balance composed principally of argon and carbon dioxide. Atmospheric air also contains water vapor, with the amount varying from negligible to several percent by weight. The air temperature and relative humidity determine actual moisture content.

The density of any fluid varies with temperature and pressure. With liquids, this variation is negligible. For gases such as air, however, the variations in density with temperature, pressure, and composition are so pronounced that even in ordinary applications they must be considered. Therefore, when dealing with airflow, the actual conditions of the air stream under question must always be defined.

In the United States, standard conditions for wastewater treatment applications are 14.7 psia (or 1 atm, 68 °F, and 36% relative humidity) (101.3 kPa, 20 °C, and 36% relative humidity). A "standard" cubic foot of air at these conditions weighs 0.075 lb (0.034 kg). The airflow rate is often expressed as standard cubic feet per minute (SCFM). Although the units appear to represent a volumetric flowrate, because of the defined properties and density SCFM actually represents a mass flowrate.

In the United States, volumetric flowrate is generally expressed as actual cubic feet per minute (ACFM) or inlet cubic feet per minute (ICFM), where ICFM refers to ACFM at the inlet to the blower. The relationship between SCFM and ACFM is:

$$ACFM = SCFM \cdot \frac{14.58}{p_b - (RH \cdot PV_a)} \cdot \frac{460 + T_a}{528} \cdot \frac{p_b}{p_a} \tag{9.1}$$

Where
  RH = ambient air relative humidity, decimal;
  PV_a = saturated vapor pressure of water at actual temperature, psi;
  T_a = actual air temperature, °F;
  p_a = actual air pressure, psia; and
  p_b = barometric pressure, psia.

Outside of the United States, volumetric flowrate is generally expressed as m³/h and mass air flowrate is often specified as normalized cubic meters per hour (Nm³/h). The specification of normalized conditions varies, but 101.3 kPa, 0 °C, and 0% relative humidity are common (14.7 psia, 32 °F, and 0% relative humidity). A normalized cubic meter of air at these conditions weighs 1.293 kg (2.85 lb):

$$\frac{m^3}{hr} = \frac{Nm^3}{hr} \cdot \frac{101.3}{p_b - (RH \cdot PV_a)} \cdot \frac{273 + T_a}{273} \cdot \frac{p_b}{p_a} \tag{9.2}$$

Where
  RH = ambient air relative humidity, decimal;
  PV_a = saturated vapor pressure of water at actual temperature, kPa;
  T_a = actual air temperature, °C;
  p_a = actual air pressure, kPa; and
  p_b = barometric pressure, kPa.

1.2   Common Blower Types

The most common blowers used in wastewater treatment can be divided into two general categories: positive displacement and centrifugal.

Positive-displacement blowers provide a fixed volume of air for every revolution of the blower shaft. The most common type consists of two counter-rotating shafts with two-lobed or three-lobed impellers on each shaft. At any given speed, positive-displacement blowers provide a constant volumetric flow rate (inlet m^3/h or ICFM) at a variable discharge pressure.

Centrifugal blowers are often referred to as "dynamic" machines. They consist of one or more impellers on a rotating shaft. The impellers use vanes that are similar to those on a centrifugal pump impeller to transfer kinetic energy to the air. The diffuser section of the blower case converts the kinetic energy (velocity pressure) to potential energy (static pressure). In contrast to positive-displacement blowers, at any given speed centrifugal blowers generally provide a wide range of flowrates over a narrow range of discharge pressure.

Centrifugal blowers are further divided into single-stage and multi-stage types. Multi-stage blowers have a sequence of impellers mounted in series along a common shaft. Single-stage blowers, as the name implies, use a single impeller to achieve the required discharge pressure.

There are other types of blowers that are occasionally used in wastewater treatment applications. These include liquid-ring, vane, and regenerative blowers. These uncommon types are generally used in low horsepower applications. The manufacturer should be contacted for details of operating characteristics and power requirements.

1.3   Blower Power Requirements

The power required for compressing air is a function of the flowrate and pressure ratio from inlet to discharge of the blower, as follows:

$$bhp = 0.01542 \cdot \frac{Q \cdot p_i \cdot X}{\eta}$$

(9.3)

Where

bhp = brake horsepower at blower;

$shaft\ Q$ = blower inlet volumetric flowrate, ICFM;

p_i = blower inlet pressure, psia;

η = blower efficiency, decimal; and

X = blower adiabatic factor:

$$X = \left(\frac{p_d}{p_i} \right)^{0.283} - 1$$

p_d = discharge pressure, psia
p_i = inlet pressure, psia

$$kW = 9.816 \cdot 10^{-4} \cdot \frac{Q \cdot p_i \cdot X}{\eta} \tag{9.4}$$

Where
   kW = power at blower shaft,
   Q = blower inlet volumetric flowrate, m^3/h,
   p_i = absolute blower inlet pressure, kPa,
   η = blower efficiency, decimal, and
   X = blower adiabatic factor:

$$X = \left(\frac{p_d}{p_i} \right)^{0.283} - 1$$

p_d = absolute discharge pressure, kPa
p_i = absolute inlet pressure, kPa

Additional energy losses for bearings, lubrication systems, and motor inefficiency should be added to the horsepower required for compression to determine the total electrical power required.

Blower efficiency is not a constant; it obviously varies with blower design. However, efficiency for a specific blower also varies with the flowrate and pressure, speed, and inlet conditions. In general, larger blowers are more efficient than smaller units of the same design. Centrifugal blowers, like centrifugal pumps, have a best efficiency point (BEP). Blower efficiency decreases the farther the actual airflow is from the BEP.

Blowers are often selected with the BEP close to the specified design airflow rate at worst-case inlet conditions. However, blowers seldom operate at either design airflow or worst-case conditions. Overall energy consumption for centrifugal blowers can usually be reduced by selecting a blower with the BEP at a lower flow than the design flowrate. As the blower airflow is reduced from design flow to match process demands, the operating point will approach and pass through the BEP. The result will be higher average efficiency over the normal operating range of the blower. To accurately determine the impact of BEP, the evaluation should be performed at average operating inlet temperature and pressure instead of worst-case design conditions.

1.4   Blower and System Curves

Blower specifications usually identify a single flowrate and discharge pressure at one inlet temperature and inlet pressure as the design point. This typically represents the worst-case operating conditions expected during system design. During normal operation, blowers actually operate at a variety of flowrates and discharge pressures. The variation of discharge pressure plotted against flowrate represents the blower's "characteristic curve," which is often referred to as the *blower curve* (see Figure 9.1). The blower power consumption is usually plotted against flowrate in a separate curve, referred to as the *power curve*. These curves are specific to a single inlet air temperature, pressure, and operating speed. The blower curves are useful in analyzing actual operating performance and determining blower power requirements.

The characteristic "curve" for a positive-displacement blower approximates a vertical line. The curve for a centrifugal blower is generally downward-sloping, with the maximum pressure at the left of the normal operating range. The shape and slope of the curve is greatly influenced by the type of vanes on the blower impeller(s). Backward curved blades provide a steep curve and radial blades provide a comparatively flat curve. Variations in speed and inlet conditions shift the curves for a given impeller.

Blowers create airflow. The system's resistance to airflow creates pressure. A blower curve by itself does not identify the actual airflow and discharge pressure, but represents a range of possible operating conditions. The discharge pressure required by the process varies with flow. The plot of this variation is called the system curve. It is similar to the system head curve for a pumping system (see Chapter 4). In most processes, the air is provided to a point below a water surface. The head of liquid above the top of diffusers or air release point creates static pressure. This is usually the largest portion of the total blower discharge pressure.

The friction from air flowing through piping, valves, and diffusers creates a pressure drop. Changes in airflow rate and control-valve position cause variations in pressure drop. Aeration systems are typically designed to allow 7 to 14 kPa (1 to 2 psi) total pressure drop at worst-case operating conditions. This includes allowance for diffusers and throttling flow control valves.

A variety of formulas and techniques are used to calculate pressure drop for specific piping configurations. These formulas are similar to those used to calculate pressure drop for the flow of water. A typical formula for clean steel pipe is seen in as follows (Compressed Air and Gas Institute, 1973):

$$\Delta P = 0.07 \cdot \frac{Q^{1.85}}{d^5 \cdot p_m} \cdot \frac{T}{528} \cdot \frac{L}{100} \tag{9.5}$$

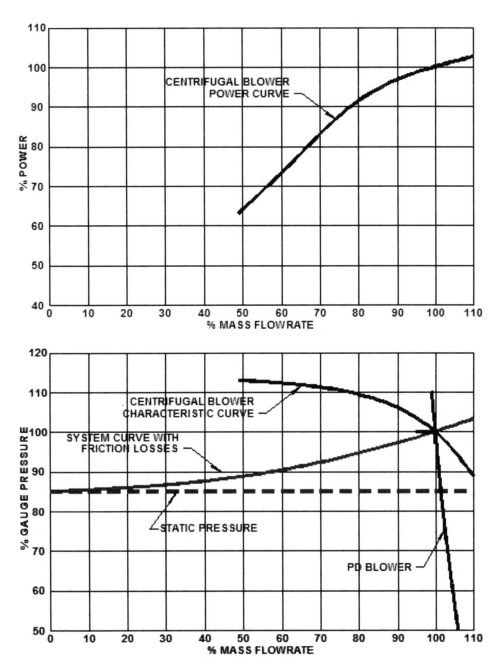

Figure 9.1   System and blower curves (courtesy of Dresser, Inc. [Dresser Roots]).

Where

ΔP = pressure drop due to friction, psi;

Q = flowrate, SCFM;

d = pipe inside diameter, in; and

p_m = mean system pressure, psia:

$$p_m = p_{initial} - \frac{\Delta P}{2}$$

T = air temperature, °R;

°R = (°F + 460); and

L = equivalent length of pipe, including allowances for fittings, ft.

In metric units:

$$\Delta P = 4 \cdot 10^7 \cdot \frac{Q^{1.85}}{d^5 \cdot p_m} \cdot \frac{T}{273} \cdot \frac{L}{100} \qquad (9.6)$$

Where

ΔP = pressure drop due to friction, kPA;

Q = flowrate, Nm³/h;

d = pipe inside diameter, mm; and

p_m = mean system pressure, kPa;

$$p_m = p_{initial} - \frac{\Delta P}{2}$$

T = air temperature, °K;

°K = (°C + 273); and

L = equivalent length of pipe, including allowances for fittings, m.

The allowance for fittings in determining the equivalent total length of pipe is obtained by adding a value for each fitting (see Table 9.1). The value will produce the same pressure drop as that length of straight pipe. These values are typically expressed in terms of equivalent diameters. For example, in 200-mm (8-in.) nominal diameter pipe, the equivalent length of each 90° elbow is 30 × 200/1000 (30 × 8/12). Each 200-mm (8-in.) 90° elbow will produce a pressure drop equal to the pressure drop in 6 m (20 ft) of straight pipe.

The relationship of pressure drop to flow through diffusers is specific to each model of diffuser and may vary over time as diffusers foul. The diffuser manufacturer should be consulted for each application. A pressure drop of 150 to 300 mm H_2O or 1.5 to 3 kPa (6 to 12 in of water or 0.2 to 0.4 psi) is common for diffusers at design airflow.

TABLE 9.1   Equivalent lengths of pipe for common fittings

Fitting Type	Equivalent Length
1.6 rad (90°) elbow	30 × diameter, m (ft)
0.8 rad (45°) elbow	16 × diameter, m (ft)
Tee, flow thru run	20 × diameter, m (ft)
Tee, flow thru branch	60 × diameter, m (ft)
Butterfly valve 100% open	20 × diameter, m (ft)
Transition	20 × diameter, m (ft)

Most aeration systems have multiple tanks operating in parallel and several grids of diffusers in each tank. By eliminating excess airflow to locations of low demand, the total system airflow can be optimized. Flow control valves are provided to control the distribution of air between the various locations. The valves at locations of low demand or reduced pressure are throttled to force the air in desired proportions to all locations. Throttling a flow control valve creates a pressure drop. The proper adjustment of the flow control valves is important to optimize energy consumption of the blowers and aeration systems. Providing proper adjustment minimizes pressure drop.

The valve's pressure drop is a function of C_v, defined as the gallons per minute of water that will pass through the valve with a pressure drop of 1.0 psi. Data on valve C_v values at various openings and sizes is readily available from manufacturers. To calculate the pressure drop caused by airflow through a valve, the following equation should be used:

$$\Delta P = \left(\frac{Q}{22.66 \cdot C_v} \right)^2 \cdot \frac{SG \cdot T_u}{P_u} \tag{9.7}$$

Where
   ΔP = pressure drop through valve, psi;
   Q = airflow rate, SCFM;
   C_v = valve flow coefficient from manufacturer's data;
   SG = specific gravity of gas, dimensionless (air = 1.0);
   T_u = upstream temperature, °R; and
   P_u = upstream pressure, psia.

Outside of the United States, the pressure drop through a valve is calculated from K_v. This is defined as the flow of water in cubic meters per hour (m³/h) that will pass through a valve with a pressure drop of 1 bar (14.5 psi). To obtain K_v, multiply C_v by 0.87, as follows:

$$\Delta P = \left(\frac{Q}{4.78 \cdot K_v} \right)^2 \cdot \frac{SG \cdot T_u}{P_u} \tag{9.8}$$

Where
  ΔP = pressure drop through valve, kPa;
  Q = airflow rate, Nm³/h;
  K_v = valve flow coefficient from manufacturer's data;
  SG = specific gravity of gas, dimensionless (air = 1.0);
  T_u = upstream temperature, °K; and
  P_u = upstream absolute pressure, kPa.

It is not generally possible to predict the actual position of flow control valves in new designs. An allowance of 3.5 to 7.0 kPa (0.5 to 1.0 psig) is usually added to the calculated pressure drop through piping and diffusers to accommodate throttling valve losses. In evaluating power consumption in existing systems, it is advisable to examine the actual status of throttling valves to verify proper adjustment and actual pressure losses.

The system curve is the sum of the static pressure and the pressure drop due to friction. Once the pressure drop for a specific set of conditions is known for one flow, the total pressure can be approximated as follows:

$$P_{total} = D \cdot 0.433 + k \cdot Q^2 \tag{9.9}$$

Where
  P_{total} = total discharge pressure, psig;
    D = depth of water at top of diffusers, ft;
    k = constant determined from calculated pressure drop at design airflow:

$$k = \frac{\Delta P_{ave}}{Q^2}$$

  Q = flowrate, SCFM
  ΔP_{ave} = average pressure drop at design airflow, psi.

In metric units:

$$P_{total} = D \cdot 9.795 + k \cdot Q^2 \tag{9.10}$$

Where

p_{total} = total discharge pressure, kPa;

D = depth of water at top of diffusers, m; and

k = constant determined from calculated pressure drop at design airflow:

$$k = \frac{\Delta P_{ave}}{Q^2}$$

Q = flowrate, Nm^3/h

ΔP_{ave} = average pressure drop at design air flow, kPa

The intersection of the system curve and blower curve identify the actual flowrate at specific conditions (see Figure 9.1).

Optimum sizing of piping and related components such as filters and valves will have a direct effect on blower power requirements. Undersized discharge piping increases friction and pressure loss, which results in increased energy requirements.

Inlet filters, silencers, piping, and control valves should be sized to create minimal pressure drop under normal operating conditions. Normal design practice is to have less than 3.5 kPa (0.5 psi) pressure drop between the air intake point and the blower inlet at design airflow. Approximately half this allowance is for pressure drop through the inlet filter, with the balance for losses in inlet pipe and fittings. Increasing filter sizes and pipe diameters can significantly reduce power consumption.

Both higher inlet pressure losses and discharge pressure requirements increase blower pressure ratio (p_d/p_i). This increases power requirements (see Eq. 9.3).

Caution should be used to avoid oversized piping. This will increase cost without providing a corresponding decrease in energy consumption. Inlet valves used for throttling centrifugal blowers and basin airflow control valves must be sized to provide adequate control while remaining within the useable range of valve travel. This is typically 15 to 75% open. Oversized valves make airflow control difficult. Oversized flow meters will not provide accurate measurement, particularly at low flowrates. Proper flow meter location, with adequate straight pipe upstream and downstream of the meter, is also critical to obtaining accurate flow measurement.

1.5   Effect of Inlet Conditions

Blower performance is greatly influenced by inlet conditions. In a typical blower application there is a wide range of temperature, pressure, and humidity encountered during normal operation. Increasing inlet temperature, decreasing inlet pressure, and increasing relative humidity decrease the density of the inlet air. A decrease in density increases the inlet volumetric flowrate the blower must provide for an equivalent mass flowrate.

It is not possible to make blanket statements on the effect of specific changes in ambient conditions on blower power consumption. The impact of the change in ambient condition varies with the type of blower, the airflow at operating conditions relative to BEP, and the blower control technique. It is also important to determine if the airflow required by the process is volumetric (m^3/h or actual cubic feet per minute [ACFM]) or mass (normalized cubic meters per hour [Nm^3/h] or SCFM).

At constant mass flowrate, an increase in inlet temperature will decrease the power required for an inlet throttled centrifugal blower, but will increase the power required for a variable-speed centrifugal blower. A positive-displacement blower at constant speed will not exhibit any appreciable power change in response to an increase in inlet temperature, but the mass flowrate will decrease. A variable-speed positive-displacement blower experiencing an increase in inlet temperature will require increased speed and power to maintain a constant mass flowrate.

Increasing inlet pressure increases air density and decreases the inlet flowrate required to maintain a constant mass flowrate for any blower. If the discharge and barometric pressure remain constant, the pressure rise across the blower is reduced and blower power is decreased.

When evaluating energy requirements and comparing alternate system designs, it is important to make sure that all evaluations are made using the same assumptions for these conditions. It is also important that the evaluation be made at typical operating conditions in order to obtain a reasonable energy evaluation. Typical operating conditions may differ significantly from specified design conditions.

1.6   Other Considerations

Ambient conditions and their influence on blower power consumption are beyond the control and influence of the designer and operator. However, a number of other design and maintenance procedures can improve energy efficiency.

An easily neglected operational consideration is proper adjustment of aeration basin airflow control valves to minimize system discharge pressure. These valves should be adjusted so that at least one valve is at the maximum open position at all times to reduce blower discharge pressure to the minimum possible value. In automatically controlled systems, this implies that "most-open-valve" (MOV) logic be incorporated into the controls. Most-open-valve logic provides automatic adjustment of a pressure setpoint or direct manipulation of valve positions to maintain at least one basin airflow control valve at or near the maximum open position. In manually controlled systems, periodic operator adjustment is required to minimize system pressure.

Blower power requirements are directly proportional to the airflow rate. Therefore, it is critical to match blower airflow to process requirements. This is not a simple matter because some processes depend on volumetric flowrate (m³/h or ACFM) for proper performance. Examples of this are channel aeration, filter backwash, and air lift pumps. Other processes depend on the correct mass flowrate of oxygen to the process (Nm³/h or SCFM). Examples of this type of process include diffused aeration and aerobic digestion.

Most treatment processes exhibit wide fluctuations in airflow demand. This is due to a number of factors. The two most significant are diurnal variations in loading and the difference between operating and design loads. Diurnal load changes occur in most municipal treatment plants. It is common for municipal treatment facilities to have a ratio of maximum to minimum daily flow of 2:1, and the minimum daily flow is typically 70% of average daily flow.

In plant design, it is normal to provide equipment sized to meet the worst-case process needs for future population and loading increases. The result is that during much or even most of a facility's life, the airflow requirements are lower than the design values. Several studies indicate that the majority of wastewater treatment facilities operate at one-third of design load. Obviously, the system design must accommodate future flows and worst-case requirements, but energy evaluations and equipment optimization should be based on realistic current flows and loads. In performing energy audits and developing conservation programs in existing facilities, evaluation of actual operating conditions is a key factor in reducing cost and consumption.

To accommodate demand variations, adequate blower turndown should be provided. Blower turndown is the ratio between maximum and minimum flowrates, as follows:

$$\text{Turndown } \% = \frac{Q_{max} - Q_{min}}{Q_{max}} \cdot 100 \qquad (9.11)$$

Where

   Q = airflow, Nm³/h (SCFM).

A total system turndown of 80% (5:1) is desirable for most applications to accommodate growth projections and diurnal variations. This turndown is usually obtained by combining changes in the number of blowers operating in parallel and taking advantage of the turndown available from individual blowers.

It is generally feasible and desirable to select blowers that can provide approximately 50% turndown. By providing 50% turndown, if one blower cannot meet the process airflow needs a second blower can be started and both blowers operated near minimum flow. This reduces the step in airflow rate and allows closer matching of the supply to the process demand, thereby minimizing blower power.

Selection of the design airflow rate for blowers while optimizing energy use is a complex task. Minimizing equipment cost and obtaining maximum design-point efficiency suggests using a few large blowers. However, the need to provide reduced airflow for extended periods of time suggests using several small blowers. In general, the energy reduction achievable with several small blowers operating over an extended period of time will offset the initial savings obtained by using a few large blowers.

Most municipal facilities require a standby blower so that 100% of design airflow is available with any one blower out of service. A good compromise between energy efficiency and equipment cost can usually be achieved by providing four blowers, each sized to provide 33% of design airflow rate. Providing four blowers with two sized to provide 50% of design airflow rate and two sized to provide 25% of design flowrate is another common configuration that provides good turndown. Using a blower configuration with good turndown will directly affect energy consumption. If operators have the ability to operate at reduced flow during plant startup, the energy use will be reduced proportionally to the airflow. If the blowers cannot provide this turndown, it will be necessary to use excess energy to operate the blowers at their minimum safe flowrate.

Another limiting factor in blower power requirements for aeration is the air flowrate needed to maintain mixing in the basins. A common recommendation for mixing airflow rate is 2.2 $m^3/h/m^2$ (0.12 CFM/sq ft) for fine-pore aeration systems. This is a conservative value. Field testing should be used to verify actual requirements. In under-loaded facilities, the air needed for mixing often exceeds the airflow rate needed to meet process oxygen demand. In this case, it would be beneficial to remove one or more basins from service. This brings mixing airflow closer to the minimum airflow needed for process demand. If it is not possible to take basins out of service, intermittent aeration can often be used to reduce average airflow. Airflow during aerated periods will provide re-suspension of solids following each non-aerated period.

Many processes, such as post-aeration or diffused aeration, operate at fixed static pressure throughout the operating flow range. The only significant change in system pressure is the variation of friction pressure drop through air-distribution piping. Other processes show a wide variation in static pressure during normal operation. Examples would include filter backwash, aerobic digesters, and sequencing batch reactors. These processes can have changes of 50% or more in diffuser submergence. In these applications, the blower system must be able to accommodate these large variations in discharge pressure. With positive-displacement blowers, the discharge pressure inherently rises and falls to match system requirements. In variable-head applications, centrifugal blowers require automatic controls to maintain desired airflow during pressure excursions.

Clearly, blower selection and energy optimization is a complex task with many interrelated variables. However, a few simple guidelines will serve to simplify design and provide direction to conservation programs in existing facilities. These are as follows:

(1) Match blower airflow to the process requirements,
(2) Minimize system discharge pressure and inlet losses, and
(3) Provide flexibility and adequate turndown in the blowers to accommodate loading variations.

These guidelines apply to all types of blowers. The details of accomplishing these goals and the limits of operation vary with the blower design.

2.0 POSITIVE-DISPLACEMENT BLOWERS

2.1 Operating Principles

In a lobe-type positive-displacement blower (see Figure 9.2), each revolution of the drive shaft causes the impeller to sweep a fixed volume of air around the outside of the case from inlet to discharge. Close clearances between the impellers and the inside of the case and between the two impellers as they mesh at the center provide a seal that prevents air from flowing backward. Gears on each shaft transmit torque from the motor and maintain the clearances between the two lobes. The motor may be directly coupled to the blower shaft or a V-belt drive may be used to operate the blower at a speed different from the motor speed.

The combined volume of air swept by the lobes during a full rotation is referred to as *blower displacement,* expressed in cubic meters per revolution or cubic feet per revolution. Positive-displacement blowers at constant speed and inlet conditions operate at constant volumetric flowrate. The impellers' seals are less than perfect.

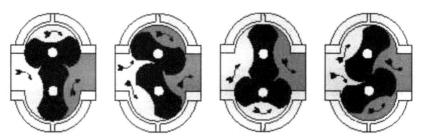

FIGURE 9.2 Positive-displacement blower operation (courtesy of Dresser, Inc. [Dresser Roots]).

This causes leakage of air from the discharge back to the inlet of the blower. This leakage is referred to as *slip* and reduces blower efficiency. To compensate for the slip, the blower must operate at a slightly higher speed (slip revolutions per minute) so that the correct total flowrate will be produced.

The positive-displacement blower airflow rate can be calculated as follows:

$$Q = Disp \cdot (N - Slip) \tag{9.12}$$

Where

Q = volumetric airflow rate, ICFM;
$Disp$ = blower displacement, cu ft/rev;
N = blower rotational speed, rpm; and
$Slip$ = slip corrected for actual operating conditions, rpm.

In metric units,

$$Q = Disp \cdot (N - Slip) \cdot 60 \tag{9.13}$$

Where

Q = volumetric airflow rate, m^3/h;
$Disp$ = blower displacement, m^3/rev;
N = blower rotational speed, rpm; and
Slip = slip corrected for actual operating conditions, rpm.

Positive-displacement blower power is composed of two parts. The *gas horsepower* represents the power to compress the air from inlet to discharge conditions. This is the largest portion of the total blower power. Additional power is required because of losses in bearings, gears, and seals. This is commonly referred to as *friction horsepower*. Friction horsepower may be treated as a constant, as follows:

$$bhp = F_g \cdot N \cdot Disp \cdot \Delta P_b + FP \tag{9.14}$$

Where

bhp = blower shaft power required, hp;

F_g = gas power constant from manufacturer (typically 0.00436);

N = blower rotational speed, rpm;

Disp = blower displacement, cu ft/rev;

ΔP_b = total pressure rise across blower, psi; and

FP = friction power corrected for actual operating conditions, hp.

In metric units, the corresponding formula is

$$kW = F_g \cdot N \cdot Disp \cdot \Delta P_b + FP \qquad (9.15)$$

Where

kW = blower shaft power required, kW;

F_g = gas power constant from manufacturer (typically 0.01647);

N = blower rotational speed, rpm;

Disp = blower displacement, m^3/rev;

ΔP_b = total pressure rise across blower, kPa; and

FP = friction power corrected for actual operating conditions, kW.

Outside of North America, positive-displacement blower airflow rates and power requirements are not commonly calculated using these formulas. Instead, flowrate and power at actual operating conditions are more commonly interpolated from tabulated values or performance curves provided by the blower manufacturer.

If the blower is belt-driven, additional power loss (typically 2 to 5%) will occur as the result of belt slippage.

2.2 Control Techniques

Positive-displacement blower airflow rate may be controlled by using unloading or blow-off valves to dump a portion of the airflow to the atmosphere. However, this technique does not reduce the power required by the blower and should be avoided. Because the blower discharge pressure will increase to match system requirements, positive-displacement blowers should *never* be throttled on the blower inlet or the discharge. Control valves at aeration basins or other process locations should only be used to distribute air proportional to the various process requirements. Adjusting these valves will not change the blower airflow rate.

The most practical control technique for positive-displacement blowers is varying blower speed. In small systems with constant speed motors, changing the blower flowrate to match average process demand may be accomplished by changing sheaves on the blower or motor. If the blower airflow is higher than process demand, changing sheaves to reduce blower speed will reduce flowrate and power requirements. If the discharge pressure is lower than design, or if the motor rating exceeds actual power draw, the blower speed can be increased to provide additional capacity. This may eliminate the need to run additional blowers at lower efficiency. When changing the blower speed, it is important to stay within the manufacturer's speed limitations and motor power rating. If several blowers are available, it may be desirable to have different operating speeds for each of them. The operator should select the blower(s) to run based on demand.

In larger systems, particularly those with automatic controls, the blower airflow should be continuously adjusted to exactly meet process requirements. A number of devices are available for this, including variable-pitch sheaves, eddy current drives, and magnetic couplings. The most common device used in current practice is the adjustable-frequency drive (AFD) (see Chapter 5). Blower airflow is directly proportional to speed. This allows manual or simple automatic controls to be used with AFDs and positive-displacement blowers.

2.3 Application Considerations

Precautions should be taken in system design to minimize the operator's exposure to noise. Inlet and discharge silencers are commonly used in piping systems. Piping in the blower room may be insulated to reduce sound radiation. Acoustic enclosures are often provided to contain noise. Some blower designs use three lobed impellers to simplify noise attenuation, although these designs may reduce energy efficiency.

The discharge of positive-displacement blowers should always be equipped with a pressure-relief valve. This prevents damage to equipment or piping resulting from excessive pressure.

2.4 Operating Limits

Air temperature rises as it is compressed in a blower. Excessive temperature rise will change internal clearances and must be avoided. Because discharge air temperature increases as speed decreases, the blower should not be operated below the manufacturer's recommended minimum speed. Motor overheating may also result if speed is reduced to a point where motor ventilation is compromised.

Blower efficiency for a specific unit will decrease as speed is reduced, so it is tempting to operate at high revolutions per minute. However, noise and mechanical stress increase significantly as speed increases. The blower must never be operated above the manufacturer's recommended maximum speed. Bearing lubrication and stress on rotating members determines the upper limit of blower speed.

As the difference between inlet and discharge pressure increases, mechanical stresses increase. This limits the allowable pressure increase across the blower. For high-pressure applications, shorter impellers are required to minimize this effect.

3.0 MULTI-STAGE CENTRIFUGAL BLOWERS

3.1 Operating Principles

The multi-stage centrifugal blower uses several impellers in series (Figure 9.3). Centrifugal force pushes air entering at the center of the impeller to the outside diameter. The flow discharged from each impeller is directed by the blower case to the inlet of the impeller in the next stage. Each successive stage increases the pressure above the previous stage. The number of stages is selected to provide the necessary discharge pressure. The last stage incorporates a volute to direct the air to the blower discharge and into the system piping.

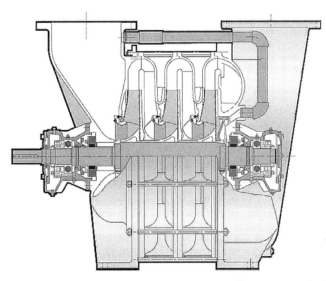

FIGURE 9.3   Multi-stage centrifugal blower (courtesy of Continental Blower, LLC/ Continental Industries).

The performance of any centrifugal blower is dependent on inlet air density. The discharge pressure that can be produced at a given airflow and the power consumed are inversely proportional to density, as follows:

$$P_a = P_c \cdot \frac{T_{ca}}{T_{inlet}} \cdot \frac{P_{inlet}}{P_{ca}} \qquad (9.16)$$

$$p_a = p_c \cdot \frac{T_{ca}}{T_{inlet}} \cdot \frac{P_{inlet}}{P_{ca}} \qquad (9.17)$$

Where

P_a, p_a = gauge pressure and power at actual conditions for a given flowrate;

P_c = curve gauge pressure at standard conditions for a given flowrate;

P_{ca} = absolute pressure corresponding to standard conditions;

T_{ca} = absolute temperature corresponding to standard conditions;

T_{inlet} = actual absolute inlet temperature;

P_{inlet} = actual absolute inlet pressure; and

p_c = curve power at standard conditions for a given flowrate.

Changes to inlet air density shift the blower's characteristic curve and power curve (see Figure 9.4). It is important to remember that these curves represent the range of possible operating flows at a specific set of inlet conditions and constant blower speed. It is necessary to identify the discharge pressure from the system curve to determine the actual airflow.

3.2 Control Techniques

The ability to vary airflow using a variety of devices is one of the advantages of centrifugal blowers. In some cases, blower airflow is controlled by discharge throttling, either intentionally or inadvertently. This will reduce airflow and power, although not as efficiently as other techniques. The techniques most commonly used for controlling multi-stage centrifugal blowers are inlet butterfly valves and AFDs.

Throttling control should be done using an inlet valve. The inlet valve serves two functions. First, it creates a pressure drop and absorbs some of the pressure rise of the blower. This moves the operating curve down relative to gauge pressure. Second, the inlet pressure drop reduces the air density, further shifting the blower curve and reducing power consumption.

The most efficient technique for controlling centrifugal blower airflow is variable-speed operation. Typically, AFDs are used to vary blower speed; however,

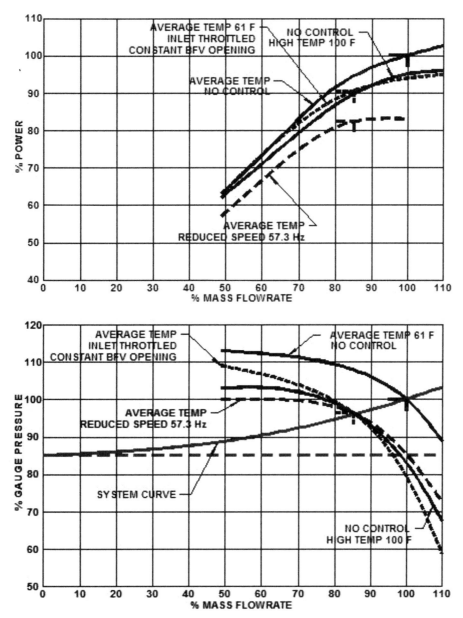

FIGURE 9.4   Centrifugal blower response to temperature and control (courtesy of Dresser, Inc. [Dresser Roots]).

magnetic couplings and internal combustion engines are also used. Variable-speed operation is 15 to 20% more efficient than throttling. The "fan laws" approximately predict the performance of centrifugal blowers operating at variable speed. When a blower is operated within normal ranges of speed variations and operating conditions the performance change can be calculated from the following:

$$Q_2 = Q_1 \cdot \left(\frac{N_2}{N_1} \right) \tag{9.18}$$

$$P_2 = P_1 \cdot \left(\frac{N_2}{N_1} \right)^2 \tag{9.19}$$

$$p_2 = p_1 \cdot \left(\frac{N_2}{N_1} \right)^3 \tag{9.20}$$

Where
Q_1, Q_2 = airflow at original and new operating speed, m³/h (ICFM);
P_1, P_2 = gauge pressure at original and new operating speed, kPA (psig);
p_1, p_2 = power at original and new operating speed, kW (hp); and
N_1, N_2 = original and new operating speed, rpm.

Not all centrifugal blowers can be controlled successfully using variable speed. If the blower curve does not have the correct characteristics, operation may be unstable. The blower manufacturer should be contacted to verify the suitability of each application.

Horizontally split multi-stage centrifugal blowers are used in high-flow applications. Inlet guide vanes (IGVs), similar to those used on single-stage blowers, are the most common control technique with this type of blower.

3.3 Application Considerations

Centrifugal blowers are sensitive to inlet conditions and discharge pressure. Occasionally, an application is encountered where the blowers will not perform properly during hot weather or periods of high airflow demand. It is much more common, however, to encounter blowers that are "over-specified." In other words, the specified design conditions far exceed the requirements of normal operation. For example, requiring maximum design flow and discharge pressure at an unrealistic inlet temperature results in an inefficient operation. As another example, efficiency is greatly reduced if the specified discharge pressure is significantly higher than normal operating pressure.

Motors for centrifugal blowers are often sized to be "non-overloading" (i.e., the motor nameplate horsepower will not be exceeded at the worst-case design conditions). If the worst-case conditions vary significantly from normal operation, this may result in operating at low motor efficiency. In special cases, it may be reasonable to allow short-term operation above motor nameplate horsepower (within the motor service factor) to improve normal operating efficiency.

3.4 Operating Limits

Centrifugal blowers are limited by the flow and pressure ranges available for a specific blower. The minimum airflow for a centrifugal blower is referred to as the surge point. An unstable pulsating flow occurs when operating below this point. The maximum pressure available from the blower usually occurs at or near the surge point. The maximum flow may be limited by motor power or system discharge pressure. Centrifugal blower operation may also be limited by the maximum discharge air temperature. This adversely affects bearing temperature and internal clearances. High-discharge air temperature most often occurs when blowers are throttled to low flows to reduce power.

4.0 SINGLE-STAGE CENTRIFUGAL BLOWERS

4.1 Operating Principles

Single-stage centrifugal blowers use one impeller to produce airflow. The single impeller operates at high speeds to provide the necessary discharge pressure.

There are two types of single-stage blowers currently available. The most common design (see Figure 9.5) uses gearing between the motor and blower shafts to increase impeller speed above motor speed. Pressure lubrication supplies oil to journal bearings on the blower and/or the gear boxes.

Another single-stage blower design has the impeller mounted directly on the shaft of a special motor operating at high speeds (see Figure 9.6). A special purpose AFD is an integral part of this design. It provides an output frequency much higher than the normal 50 Hz or 60 Hz supplied by utilities. This gearless single-stage blower is often referred to as a high speed turbo blower. Most designs use special bearings to support the blower and motor shaft and do not require external lubrication systems for the bearings.

The operating principles of single-stage and multi-stage centrifugal blowers are similar. Both exhibit variation in operating characteristics with changing inlet air

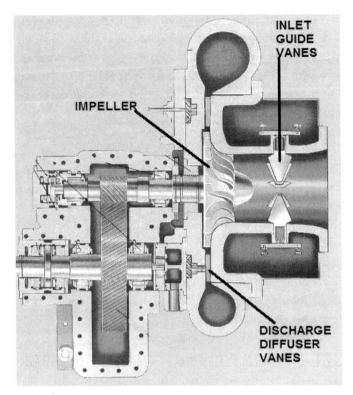

FIGURE 9.5   Single-stage centrifugal with gearing and guide vanes (courtesy of
Dresser, Inc. [Dresser Roots]).

density and speed. Single-stage blowers are typically higher in cost than multi-stage
blowers, but have higher operating efficiencies.

4.2   Control Techniques

The airflow rate of single-stage centrifugal blowers may be varied using guide vanes
(see Figure 9.7), throttling, or variable speed.

Modulating the airflow of high-speed turbo blowers is achieved by varying the
speed of the blower with the integral AFD. Most high-speed turbo blower systems
include integral control systems to modulate flow and provide surge protection.

Varying blower speed is the most efficient method for varying airflow (Moore,
1989). Geared single-stage blowers could be controlled using variable speed. However, in the past this type of blower was commonly provided with medium voltage

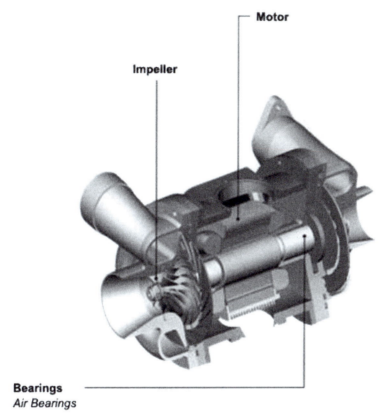

Motor

Impeller

Bearings
Air Bearings

FIGURE 9.6   High-speed turbo blower (courtesy of HSI, Inc.).

motors (>600 Volts alternating current [VAC]). Medium-voltage AFDs were not used because they were either not available or were extremely expensive. Capacity control and power reduction were typically provided by adjustable IGVs or variable discharge diffuser vanes (DDVs).   *Inlet Guide Vanes*

An IGV swirls the inlet air in the direction of impeller rotation prior to the air entering the eye of the impeller. This changes the "head" of the blower and results in reduced discharge pressure and lower energy requirements at a given air flow rate (Figure 9.8). Reducing the IGV opening effectively shifts the blower characteristic curve down and to the left.

A DDV (see Figure 9.9) changes the conversion of the kinetic energy of the air exiting the impeller to potential energy. As the DDV is closed, the blower curve is

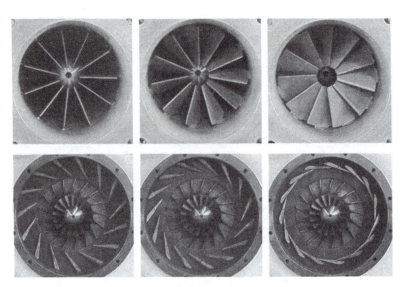

FIGURE 9.7   Inlet Guide Vanes (top) and Discharge Diffuser Vanes (bottom) (courtesy of Dresser, Inc. [Dresser Roots]).

shifted to the left, reducing flowrate and power consumption at a given discharge pressure. Some blowers use both IGV and DDV control. They use proprietary algorithms to coordinate the two types of vanes and optimize efficiency.

Medium-voltage (>600 VAC) AFDs are becoming more common, and models are available in lower horsepower ratings. This makes the application of AFDs to geared single-stage centrifugal blowers economically feasible for energy reduction in some cases. A careful analysis is required, and the blower manufacturer should be consulted to determine the required operating parameters.

Throttling can also be used to provide flow control for single-stage blowers. However, this is not as efficient as other control techniques.

4.3 Application Considerations

Application considerations for multi-stage centrifugal blowers also apply to single-stage centrifugal blowers. Single-stage blowers are more sensitive to surge than multi-stage blowers. A blow-off valve is usually provided to prevent surge during startup. Opening of the blow-off valve should be minimized to avoid wasting power.

Single-stage blowers may generate high noise levels because of the high impeller tip velocity. Sound containment enclosures and silencers on inlet, discharge, and blow-off piping are typically used to minimize the exposure of operators to noise.

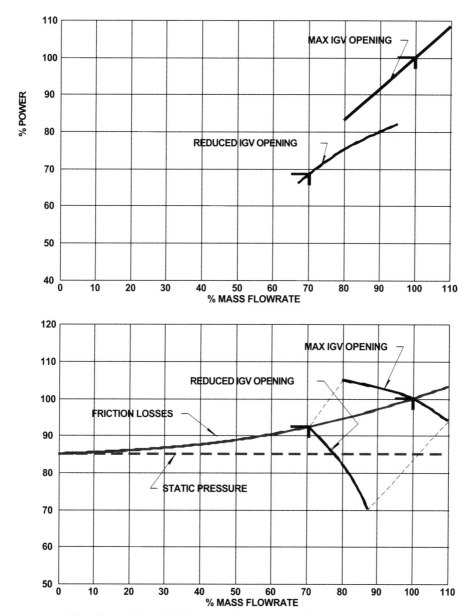

FIGURE 9.8   Inlet Guide Vane (IGV) control (courtesy of Dresser, Inc. [Dresser Roots]).

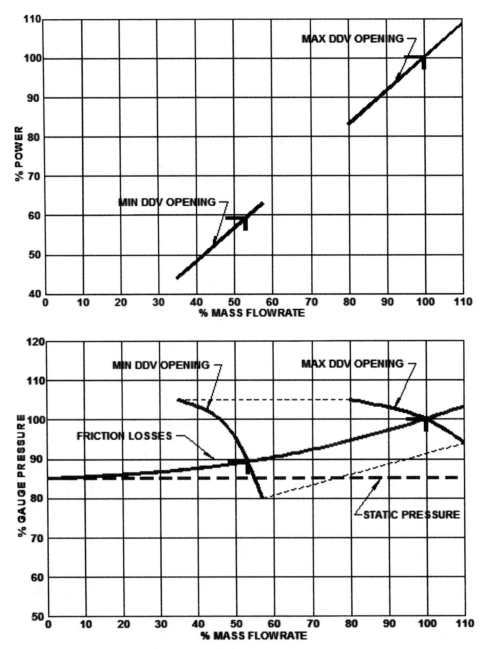

FIGURE 9.9 Discharge Diffuser Vane (DDV) control (courtesy of Dresser, Inc. [Dresser Roots]).

The power required for enclosure ventilation should be included when evaluating blower power consumption.

4.4   Operating Limits

Single-stage centrifugal blowers exhibit the same flow and pressure range limitations as multi-stage centrifugal blowers.

5.0   BLOWER SYSTEM RETROFIT OPPORTUNITIES

5.1   Automatic Controls

The variation in aeration system demand makes automatic control of blower airflow cost-effective in most applications. Significant energy reductions are achievable by matching blower airflow to process requirements. Equal or even improved process performance may be obtained.

Automation should begin with control of individual blowers. Blower controls can incorporate blower monitoring and protection functions with capacity control functions. This allows the operator to conveniently adjust airflows and permits operation closer to blower limits without jeopardizing equipment life. With centrifugal blowers, the controls can maintain constant airflow by compensating for fluctuations in ambient or discharge conditions. Proper application of automatic controls can also allow the use of centrifugal blowers in variable-pressure applications such as aerobic digesters and sequencing batch reactors. With positive-displacement blowers, adding AFDs and controls will allow the airflow rate to be changed frequently and conveniently to match requirements.

A common application for blower controls is maintaining a constant discharge pressure. This is used to prevent interaction between the flow controls of multiple aeration basins. However, selection of an excessively high pressure setpoint or failure to properly adjust the basin flow control valves to minimize pressure can lead to wasted power. Therefore, care must be exercised in adjusting this type of system.

Using automatic control of dissolved oxygen concentration can reduce blower power consumption by 25 to 50% compared to manually controlled systems. The objective in dissolved oxygen control systems is to match airflow to process demand. This is accomplished by maintaining a constant dissolved oxygen concentration. In applications using small blowers (generally less than 110 kW [150 hp]), it is common to control total blower airflow automatically and have the airflow distribution between multiple tanks controlled manually. In larger systems, it is cost-effective to also provide automatic flow control for each aeration tank. This reduces the variations in dissolved oxygen concentration between tanks. In large systems, it may be cost-effective to control even individual zones or diffuser grids within each tank. This allows continuous optimization of the dissolved oxygen concentration throughout the tank, which minimizes energy consumption.

In all cases of automatic dissolved oxygen control, aeration basin logic and instrumentation should be integrated with the blower controls. If automatic control of airflow to individual tanks or zones is provided, MOV logic should be used to minimize system pressure.   *Most Open Valve*

5.2 Additional Energy Conservation Measures

Premium efficiency motors should be used with blowers because of their significance to total plant energy consumption. Older blower motors may have low efficiency and it will generally be cost-effective to replace them with high-efficiency motors (see Chapter 3).

When considering motor replacement strategies, it may not be advisable to match the size of the new motor to the existing one. As previously indicated, motors are often sized to meet worst-case design conditions. Operating data should be collected over a range of actual conditions. Improvements in efficiency may be obtained by providing a motor rated closer to actual requirements.

The addition of capacitors to constant-speed blower motors will usually provide cost-effective improvement if power factor penalties are assessed by the electric utility (see Chapter 2). When applying power factor correction capacitors to reduced-voltage solid-state starters, the starter manufacturer's recommendations should be followed to avoid damage to the starters.

For positive-displacement blowers, a simple sheave change may be all that is required to achieve improvement in efficiency by matching flowrate to demand.

Preventive maintenance procedures can have an impact on energy consumption as well as equipment life. Belt drives should be properly tensioned. Excessive tension reduces bearing life and increases bearing friction. Insufficient belt tension increases belt slip, which wastes power and reduces belt life. Proper lubrication of motors and blowers will reduce energy consumption. It is important to note that over-lubrication will decrease both efficiency and equipment life.

Inlet filters should be cleaned regularly and monitored for excessive pressure drop. Clogged filters increase pressure drop, which results in greater energy consumption. In many installations, inlet filter removal efficiency is over-specified (U.S. EPA, 1989). This results in extreme pressure drop. Field experience has demonstrated that inlet filters capturing 95% of particles 10 micron (0.000039 in.) or larger are adequate for most fine-pore aeration systems. They also meet the requirements of most blower manufacturers. By using replacement cartridges or filters with this removal efficiency or higher flow capacity, reduced energy consumption can be achieved. This will not affect blower or diffuser life.

Modification of the blowers themselves may be a cost-effective way to improve the efficiency of centrifugal blowers. Substantial improvement in energy performance can be achieved by changing impellers to more closely match actual operation. This is particularly true when actual discharge pressure during normal operation is lower than design values. In extreme cases, especially if installed blowers cannot provide adequate turndown, it is cost-effective to completely replace some blowers with smaller units. Table 9.2 presents examples of energy conservation measures.

TABLE 9.2   Examples of energy conservation measures

Description	Comments
Reduce inlet filter losses	Replace existing filter elements with less restrictive ones. Provide regular preventive maintenance to replace dirty elements with clean ones.
Replace older motors with high-efficiency units	Verify that motor size is matched to actual load.
Change sheaves to increase or decrease flowrate (positive-displacement blower only)	Match blower output to actual demand. Verify motor rating will accommodate new flowrate.
Trim or replace impeller (centrifugal blower only)	Match blower output to actual demand and pressure. Verify that the combination of new flow and pressure will result in lower energy consumption.
Provide AFDs to modulate airflow	Verify motor insulation (Class F or better) and proper grounding. Verify centrifugal blower curve will accommodate variable-speed control.
Use automatic blower control to regulate blower airflow	Match flow to load more accurately.
Use automatic dissolved oxygen control to regulate blowers	Continuously match blower flowrate to aeration demand.
Add MOV logic to aeration controls	Reduce discharge pressure to minimum requirements.
Improve blower and motor maintenance	Check belt tightness, verify lubrication, change filters regularly.
Install smaller blowers or blowers with more turndown	Match blower output more closely to process requirements.

6.0 REFERENCES

Compressed Air and Gas Institute (1973) *Compressed Air and Gas Handbook;* Compressed Air and Gas Institute: Cleveland, Ohio.

Dresser Roots (2004) *Product Data Book;* Dresser Inc., Dresser Roots: Connersville, Indiana.

Gartmann, H., Ed. (1970) *De Laval Engineering Handbook;* McGraw-Hill: New York.

Cantwell, J. (2007) Value of Energy Savings at Small Facilities. *Proceedings of the 80th Annual Water Environment Federation Technical Exhibition and Conference;* San Diego, California, Oct 13-17; Water Environment Federation: Alexandria, Virginia.

Moore, R. L. (1989) *Control of Centrifugal Compressors;* Instrument Society of America: Research Triangle Park, North Carolina.

U.S. Environmental Protection Agency (1985) *Summary Report, Fine Pore (Fine Bubble) Aeration Systems;* U.S. Environmental Protection Agency, Water Engineering Research Laboratory: Cincinnati, Ohio.

U.S. Environmental Protection Agency (1989) *Design Manual Fine Pore Aeration Systems;* EPA/625/1-89/023; U.S. Environmental Protection Agency, Office of Research and Development: Cincinnati, Ohio.

Chapter 10

Solids Processes

(continued)

1.0 INTRODUCTION

Wastewater sludge is the residuals left from the treatment of wastewater. Sludge is usually categorized by the process from which it is removed (e.g., primary sludge, waste activated sludge, or digested sludge). Sludges removed by similar processes often have similar consistency, inorganic content, and dewaterability, but sludges removed from dissimilar processes typically have differing characteristics. From an energy perspective, the two most important characteristics of sludge are its water content and the enthalpy or heat content of its dry matter.

Water is difficult to remove from sludge and, beyond separation by simple settling (gravity thickening), typically requires chemicals or energy for further removal. The cost of removal of water increases exponentially with the water content of the sludge product desired (i.e., the cost of drying is greater than the cost of dewatering, which is greater than the cost of thickening). Organic matter present in the sludge represents unburned fuel and, therefore, has inherent energy or heat value. The wastewater industry often refers to volatile suspended solids (VSS) as representing organic matter and, indirectly, heat content. In most heat calculation examples found in the literature, it would appear that the ratio of heat content to VSS was constant at 23 000 kJ/kg VSS (10 000 Btu/lb). In reality, the heat-content-to-VSS ratio will vary with the type of sludge and the degree of decomposition already accomplished; that is, digested sludges have lower ratios of heat content to VSS than non-digested biological sludges and biological sludges have lower ratios than raw sludges.

1.1 Recycle Streams

Minimization of contaminants in any sludge recycle stream is an important goal. Recycle streams not only return solids and biochemical oxygen demand back to wastewater treatment processes, but also unwanted nitrogen, phosphorus, and organic acids. Concentrated recycle streams can reduce the treatment capacity of a wastewater treatment plant (WWTP), cause process upsets, and result in needless energy waste. Understanding the effect of recycle streams on treatment processes and knowing how to control them are essential to proper process control. Numerous books and technical papers have addressed these important issues.

Gravity thickeners should be operated to provide optimal concentration of solids without allowing needless aging of the sludge or excessive loss or washout of solids in the recycle stream. Dewatering units and chemical belt and drum thickeners should be operated to provide for a high solids capture rate while achieving maximum solids concentration and dewatering. Capture rates of 90% or more are achievable and should be targeted even at the expense of slightly less-thickened solids.

1.1.1 Energy-Saving Opportunities in the Treatment of Recycle Flows

Recent advances in our understanding of the microbiology of the nitrogen cycle have allowed the development of energy-efficient biological treatment process for nitrogen removal from recycle flows. Recycle flows from dewatering after anaerobic digestion have high concentrations of ammonia (1000 to 1500 mg $N-NH_3/L$) that would impose a significant aeration demand at the treatment plant and could also require the addition of pH control chemicals and/or an external carbon source for denitrification. The conventional way of conducting nitrogen removal is by achieving conversion of ammonia into nitrates in the nitrification process and then converting the nitrates into nitrogen gas in the denitrification process. The equations that summarize the reactions are as follows:

$$NH_4^+ + 1.5\ O_2 = NO_2^- + H_2O + 2H^+ \qquad (10.1)$$

$$\left.\begin{array}{c} \\ \\ \end{array}\right\} \text{Nitrification}$$

$$NO_2^- + 0.5\ O_2 = NO_3^- \qquad (10.2)$$

$$NO_3^- + 5\ \text{methanol} + CO_2 = 3\ N_2 + 6\ HCO_3^* + 7\ H_2O\ \text{Denitrification} \qquad (10.3)$$

According to these equations, the oxygen requirements for nitrification are 4570 g of oxygen/kg of ammonia-nitrogen (4.57 lb/lb) and 2000 g of methanol/kg of nitrate (2 lb/lb). A new process has been developed to denitrify using nitrites, NO_2^-, by avoiding the second part of the nitrification reaction from nitrites to nitrates. The oxygen requirements are reduced by 25% to 3400 g of oxygen/kg of ammonia-nitrogen (3.4 lb/lb). In addition, there are also savings of 40% in the denitrification carbon demand according to the following equation:

$$NO_2^- + 3\,\text{methanol} + 3\,CO_2 = 3\,N2 + 6\,HCO_3^- + 3\,H2O + \text{bacterial growth} \quad (10.4)$$

Several plants are fully operational in Europe using this principle. The process can be further optimized by using ammonia instead of methanol to drive the denitrification step according to the following equation:

$$NO_2^- + NH_4^+ = N_2 + 2\,H_2O \quad (10.5)$$

This reaction is known as an anaerobic ammonia oxidation (or ANAMMOX) reaction, and the bacteria responsible for performing the reaction are of the genus *planctomyces*. Recycle flows from solids processing operations are good candidates to use with this process because of the high concentration of ammonia and the relatively low carbon/nitrogen ratio. By combining this process with partial nitrification, 50% to nitrites (partial nitritation), and then using the other 50% of the ammonia for denitrification, it is possible to achieve full nitrogen removal without using a carbon source and reducing the oxygen requirement by 60%. Only 1700 g of oxygen/kg of ammonia-nitrogen (1.7 lb/lb) is required. In reality, there is some production of nitrate and the requirements for carbon are reduced by 90% as opposed to 100% in the aforementioned equations.

A full-scale installation for the treatment of return flows from dewatering operations has been functional in Austria since 2004 using a sequencing batch reactor. In this case, the partial nitrification and anaerobic ammonia oxidation reactions are induced by carefully controlling the reactor pH and aeration cycles. The measured reductions in energy use for ammonia removal at the Austrian plant confirm 60% theoretical energy reduction and reduction in carbon source for denitrification. A pilot study of this process has been conducted in the United States by the District of Columbia Water and Sewer Authority. A reported disadvantage of these processes is the long startup times due to the slow growth rate of the microorganisms involved.

1.2   Process Removals

In general, it is best to optimize the primary treatment process to save money and energy downstream. Care must be taken, however, to avoid too great a primary sludge age caused by either using too many primary clarifiers for the given flow or too deep a primary clarifier sludge blanket. This is especially important if co-settling with waste biological sludge is practiced. If organic sludge containing large populations of microorganisms is held for long periods, the first stage of anaerobic decomposition will begin to occur—acid fermentation with resultant drop in pH, reduction in alkalinity, and gas bubbling through release of carbon dioxide. Such conditions are typically detrimental to the performance of downstream biological processes.

Use of chemicals to improve removals in the primary clarifier can be of value. Chemically enhanced primary treatment can double the amount of total suspended solids (TSS) and biochemical oxygen demand removal in primary clarifiers, reducing the organic load to secondary treatment processes and, consequently, the energy requirements for aeration. The tradeoff is in savings from less electrical energy used in downstream processes, increased gas production in anaerobic digesters, and, possibly, lower dewatering chemical cost and drier cake solids versus the cost of the chemical and any increase in costs associated with an increase of sludge as a result of the chemical itself.

1.3 Thickening Optimization

The importance of thickening in downstream stabilization processes cannot be overemphasized. Pre-thickening before stabilization processes reduces the volume of sludge to be processed, increasing the solids retention times (SRTs) in the digesters. Digester performance in both aerobic and anaerobic systems is dependent on SRT. Longer SRTs provide enhanced stabilization and pathogen density reduction. In addition to improved performance, thickening of sludges is important for digestion because the process

- Decreases the mass of sludge to heat in anaerobic digestion or autothermal thermophilic aerobic digestion (ATAD). Heating of incoming sludge is the single most important use of energy in anaerobic digesters;

- Allows auto-heating of the digester contents in ATAD systems;

- Affects the total solids concentration the digester operates at, affecting mixing processes and oxygen transfer processes; and

- Reduces the volume of side-stream return flow after digestion, decreasing hydraulic loadings.

2.0 ANAEROBIC DIGESTION PROCESSES

In the anaerobic digestion process, organic matter is broken down to methane, carbon dioxide, ammonia, and water. This is a natural and relatively energy-efficient process. When properly applied, anaerobic digestion can result in a net production of energy, as opposed to alternative stabilization methods, which are essentially all net consumers of energy. Despite this, anaerobic processes are often not used in applications where they are favorable, and many facilities that use anaerobic treatment do not use the energy advantages of these processes to their full potential. An additional advantage of a digestion process is reducing the mass of solids to be dewatered and ultimately hauled from the site, achieving secondary energy savings.

Anaerobic decomposition will occur at all temperatures between freezing and almost boiling, but optimum production of methane only occurs in the 32 to 35 °C (90 to 95 °F) mesophilic range and in the 54 to 57 °C (130 to 135 °F) thermophilic range. Methane-rich digester gas can be collected and burned to produce energy and/or heat. Digester gas can also be upgraded to natural gas quality that can be used in the gas grid or compressed to produce vehicle fuel. Because the conventional digestion process must be operated at approximately 35 °C (95 °F), the process itself requires a small amount of heat input as well as other energy input for mixing or recirculation. With the exception of cold winter periods in northern climates, properly designed and operated digesters should be net energy producers. Digester gas can be burned in engines to drive equipment or generate electricity. There is typically sufficient waste heat, which can be recovered from the engines to provide the necessary heat back to the digesters to maintain mesophilic temperatures. In addition to heating digesters, the waste heat can also be used to provide facility heating. Table 10.1 compares some of the advantages and disadvantages of anaerobic treatment.

TABLE 10.1 Advantages and disadvantages of anaerobic stabilization processes

Advantages	Disadvantages
High degree of stabilization; inactivates pathogens	Slow growth rate of methanogens
	Requires long solids retention times
Lessens amount of sludge for final disposal	May require auxiliary heating
	Capital intensive
Low nutrient requirements	Maintenance intensive
Low energy requirements	Generates a poor-quality side stream
Methane-rich gas is usable product	Methane is a powerful greenhouse gas that requires collection
Stabilized sludge is usable product	Biogas is usually odorous

2.1 Temperature

Anaerobic processes are typically operated in the mesophilic range, 32 to 38°C (90 to 100°F), although operation at higher temperatures in the thermophilic range (50 to 60°C [122 to 140°F]) is increasingly popular. Most of the agencies that operate thermophilic anaerobic digesters use a temperature of 54 to 55 °C (130 to 131 °F). Stable operation of thermophilic digesters has been reported at this temperature. The ability to obtain consistent temperature is key to the success of thermophilic anaerobic digestion. Thermophilic anaerobic bacteria appear to be especially sensitive to temperature variations. The requirement for consistent thermophilic temperature implies that adequate mixing within the digester is also of importance to avoid the formation of pockets of sludge at reduced or increased temperature.

It is now well accepted that thermophilic anaerobic digestion will destroy more volatile solids than mesophilic digestion systems at the same total SRT. The amount of additional volatile solids reduction (VSR) depends on several factors. At relatively low SRT (15 to 20 days), the thermophilic system is likely to achieve about 4 to 8 percentage points of additional VSR. For example, if the mesophilic system achieved 50% VSR, the thermophilic system at the same SRT is likely to achieve between 54 and 58% VSR. At high SRT (30 days), there will tend to be less difference between the two systems (Willis and Schafer, 2006). Alternatively, the same VSR amount can be achieved in shorter total SRT and, consequently, a smaller tank, when using thermophilic systems instead of mesophilic systems.

Thermophilic anaerobic digestion requires additional heating to achieve the operational temperatures required. An energy balance comparison for a conventional mesophilic and thermophilic digester is presented in Figure 10.1. It is observed that during the winter period the energy demand for heating the feedstock solids in the thermophilic digester is almost twice the amount required for a conventional mesophilic digester. Energy requirements for heating the digester feedstock represent the principal energy demand for a digester. Thermophilic digestion increases this heat demand significantly. Heat loss in the digester to maintain thermophilic temperature is balanced out by the smaller tank sizes, and there is not a significant difference. An increase in the amount of biogas produced during thermophilic digestion is not enough to balance out the increased demands associated with feedstock heating. To minimize heat input, designers and operators usually provide a thicker feedstock. It is common to thicken the sludge to at least 5% solids and, in most cases, to 6% and higher. Others recover heat of the discharged biosolids to preheat the raw sludge feedstock and reduce the total fuel required.

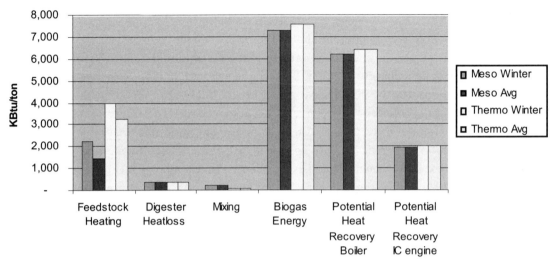

FIGURE 10.1 Energy balance comparison of mesophilic and thermophilic anaerobic digestion (4% solids).

One disadvantage of thermophilic anaerobic digestion is the higher amount of residual volatile fatty acids in the digested solids. It is common to follow the first thermophilic anaerobic digester with a series of secondary mesophilic ones to provide polishing and a high-quality final product. This type of process is called *temperature-phased anaerobic digestion* (TPAD) and has been reported to produce a higher VSR and better pathogen destruction than conventional mesophilic digestion. From an energy perspective, a tradeoff exists between the increased gas production associated with higher VSR for similar tank volumes when using TPAD and the additional heat required to run a thermophilic process. The reported gas yield for TPAD processes, 0.9m³/kg (15 cu ft/lb) of volatile solids destroyed, is similar to other anaerobic digestion process for municipal sludge (Krugel et al., 2006). The result of a life cycle cost analysis comparison is usually dependent on the local conditions.

2.1.1 Class A—Anaerobic Digestion

Thermophilic anaerobic digestion has been considered by utilities and engineers to produce Class A biosolids from municipal wastewater sludges. As thermophilic anaerobic digestion is not listed in U.S. CFR Part 503 regulations as a process to further reduce pathogens, the ability of these systems to produce biosolids meeting Class A pathogen requirements as defined in U.S. Environmental Protection Agency's

(U.S. EPA's) 40 CFR Part 503 regulations has been addressed on an application by application basis (U.S. EPA, 2009). A variety of different reactor configurations using combinations of thermophilic and mesophilic digesters have been evaluated recently (Willis et al., 2005).

2.2 Staged Anaerobic Digestion

An alternative to conventional single-stage mesophilic digestion is staged anaerobic digestion. In two-stage digestion, reactors are arranged in series rather than in parallel, which has configuration advantages regarding VSR and pathogen reduction. Several recent examples have confirmed the good performance of staged digestion. Reactors are usually operated in a draw/fill schedule (pumped transfer followed by refilling) to minimize the effects of short-circuiting. Operated in a continuous-feed mode, the use of staged digestion can improve operational performance and stability.

2.3 Two-Phase Digestion

Two-phase digestion is a type of staged digestion where the two principal steps of the anaerobic decomposition of sludges, acid formation and methanogenesis, are separated into two different digesters. This separation allows for a better performance of each of the individual phases. In two-phase digestion, the first digester is highly loaded, with a target load of 32 kg $VSS/m^3 \cdot d$ (2 lb/d/cu ft), with low SRTs of 1 to 2 days. This digester is called the *acid-phase digester*. The methanogenesis phase, also denominated gas phase, is carried out in a series of digesters that receive the fermented sludge from the acid digester. Side by side comparison of the performance of two-phase digestion has indicated that the process consistently achieves a 10% increase in VSR with respect to conventional mesophilic digestion, digesting a mixture of primary and secondary sludge at total SRTs of 12 days in the two-phase system. The reported overall gas yield of the two-phase process is similar to the gas yield of a conventional digestion process, with a reported value of 0.94 m^3 gas/kg volatile solids (15 cu ft/lb) destroyed. The composition of the gas in the acid-phase digester is one-third methane and two-thirds carbon dioxide, which is almost the reverse of the conventional digestion process. However, the volume of biogas produced in the acid digester is less than 10% of the total biogas yield of the combined process. The composition of the gas in the gas digester shows increased methane concentrations with average values of 65% methane. Maintaining control of the loading/SRT of the acid-phase digester is critical to the process.

2.4 Torpey Process

In the Torpey process, a portion of the digested sludge is mixed with feedstock sludge and returned to the digesters ahead of the thickening step in a similar way as the activated sludge is returned to the aeration basins after secondary clarification. The process was first tested in the 1960s in New York City. More recently, a modification of the process has been applied by using a dissolved air flotation system to thicken a portion of the digested sludge prior to returning it to the head of the digester. This practice effectively increases the sludge residence time in the digester without affecting the hydraulic residence time. Alternatively, a higher volumetric loading rate can be achieved without affecting the SRT of the process. Applications of the Torpey process have exhibited a 10 to 15% increase in VSR as well as an increase in the biogas yield. The original report from the application of the process in New York City indicates a 20% reduction in the total amount of dry solids to be disposed of when using gravity thickeners.

2.5 Co-Digestion Processes

Co-digestion refers to the augmentation of the amount of digestible organic matter that is introduced to the digester to enhance methane production by using sources other than plant sludge. Typical substrates tested for co-digestion are

- Fats, oils, and greases (FOG);
- Food waste;
- Organic fraction of municipal solid waste (OFMSW); and
- Waste organic matter from industrial processes.

Table 10.2 presents a summary of recent studies with co-digestion of WWTP sludge and other substrates. As the table shows, co-digestion has been successfully tested in full-scale applications for augmenting the methane production rate in digesters. It is necessary to pay attention to pretreatment of the cosubstrate to minimize operational problems with the digestion process and digester equipment such as mixers, pumps, and heat exchangers. Substantial pretreatment is required for successful processing of food wastes and OFMSW.

One of the main concerns associated with co-digestion is the increase in concentration of total solids within the digester, as the aforementioned substrates contain high solids concentrations. This is particularly true with OFMSW, which has a relatively low biodegradability. After converting the biodegradable organic matter to

TABLE 10.2 Summary of information on co-digestion studies

Item	Units	Treviso WWTP	East Bay-Oakland	Frutigen	Riverside, CA
			Case Study		
Sludge		WAS	None	Primary + WAS	Primary + WAS
Co-Substrate		OFMSW	Food Waste	Food Waste	Grease
Pretreatment		Metal and Plastic Removal, Shredding, Grit removal, Flotable removal	Dilution, Grinding,Grit removal, pulping,	Metal and Plastic Removal, Shredding, Grit removal, maceration	Screening,heated storage, mixing,
Influent TSS	%	4	4 to 10	6.5	1.2-12
Digester TSS	%	3	2 to 3	3.3	
VSR	%		50 to 80		
Digeser Loading	lb VS/ft3 d	0.048	0.2-0.6		
Co-Substrate load	%	25	100	20	30
Biogas Rate Increase	%	360	N.A.	27	90
Biogas Yield cosubstrate	ft3/lb VS applied d	12.5	7	9.2	13
Co-Substrate TVS	%TSS	67	90	91	95
Digester TVS		56	65		
Biosolids produced	lb/lb		0.28		Decreased by 30%
Biogas production rate	ft3/ft3 d	0.34	4	0.95	
Biogas yield	ft3/lb VS applied	7.0	7.0	9.2	13

methane, a significant portion of biologically inert solids still remain in the digester. Viscosity of sludges increases with total solids concentration and could affect the efficiency of mixing and pumping of the digester contents. For this reason, the total solids concentration of the mixed substrate introduced to the digester is maintained relatively low (4% in most instances), as seen in Table 10.2. Grease waste is less susceptible to generating increased total solids in the digester because it is highly biodegradable (95% reported) and would not leave significant biologically inert total solids. However, an effective mixing and recirculation system is necessary when codigesting FOG to avoid formation of a "scum" layer that can result in significant operational problems. Pulping of food wastes to remove low biodegradable solids has been reported to increase biodegradability of the substrate to 80%, minimizing the effect of solids accumulation. Evaluation of mixing capacity in the digester is recommended before embarking on a co-digestion project.

The inert solids not gasified in the digester will add to the total amount of solids to be dewatered and finally disposed of from the facility. Full-cost evaluation of the effects of added substrate is then necessary to evaluate the merit of co-digestion.

The addition of cosubstrates has been reported to improve the VSR of the plant sludge in the digester. This further complicates the evaluation because total solids are increased due to the inert residue of the cosubstrate, but total solids are decreased because increased biodegradability of the plant sludge is induced by the cosubstrate. When using FOG as a codigestion material, there is at least one report where the net effect is a reduction in the total amount of final solids to be disposed of after co-digestion.

The effect of adding the cosubstrate on the methane generation rate seems to be positive in all cases. Biogas generation rates increase from 30 to 300%. The latter number was obtained in a plant with significant excess digestion capacity digesting only waste activated sludge (WAS). Grease addition has been reported to almost double the gas production capacity of the digester with addition of grease in a proportion of 30%, by weight. Not only is grease highly biodegradable, but its methane yield is twice that of combined primary and WAS sludge or up to three times the yield of WAS alone.

2.6   Pretreatment of Sludge

There has been considerable amount of interest recently in the development of technologies that would pretreat excess sludge from municipal treatment plants to increase its biodegradability or degradation rates or both. Several technologies have been tested in laboratory-, pilot-, and full-scale applications for disintegration of WAS. The theory behind it is that the limiting step in the biodegradation of the bacterial cell is the lysis of the cell wall and membrane. It is expected that by breaking down the cell and releasing the protoplasm contents, the biodegradability as well as the rate of biodegradation would increase. An increase in biodegradability of the sludge would imply a reduction, or minimization, of sludge production, while an increase in the degradation rates might result in reduced tanks for digestion. Several physical as well as chemical treatments of WAS or portion of the return activated sludge (RAS) stream have been tested. These include the following:

- Ozonation,
- Ultrasound homogenization,
- Shear milling,
- Thermal hydrolysis,
- Focused-pulse treatment,
- Chemical-pressure pretreatment, and
- Enzymatic hydrolysis.

If the treatment is located in a portion of the RAS stream, then the emphasis is on minimizing the amount of WAS. If the treatment is located in the WAS stream before digestion, then the emphasis is not only on reducing the amount of sludge, but also on increasing the performance of the digestion step. The following aspects are important to consider when evaluating implementation of these technologies:

1. *Energy expenditures to run the disintegration process*—Energy expenditures between 0.1 and 1 kWh/lb TSS have been reported for different disintegration processes. In some of the thermal hydrolysis processes, there are important energy expenditures for running the steam boilers and pulping equipment.

2. *Ammonia and organic nitrogen release from bacterial cell disintegration*—Approximately 1% of the mixed liquor suspended solids (MLSS) is organic nitrogen that would be solubilized with cell disintegration. This organic nitrogen will add to the oxygen demand of the plant through nitrification.

3. *Additional degradable chemical oxygen demand from disintegration*—When the disintegration process is located in the RAS line, the solubilized degradable organic compounds released during disintegration will provide a source of carbon for denitrification or an additional oxygen demand in the aeration tank. The location of the unit, before or after thickening, is important when evaluating this impact.

4. *Energy gains in the anaerobic digestion process*—When located upstream of an anaerobic digester, a disintegration process changes the kinetics of degradation, increases volatile solids destruction, and augments the biogas yield. However, recent full-scale tests with one of the processes indicate that an increase in VSR does not proportionally increase the methane yield. Disintegration affected the kinetics of degradation, but not the methane yield significantly. No generalizations can be made here and each process seems to have a different effect on anaerobic digestion that requires individual evaluation.

5. *Viscosity changes*—A significant change in the viscosity of the disintegrated WAS has been reported. Improvements in the mixing of the anaerobic digestion process can be expected with lower viscosity.

6. *Cost reduction in downstream sludge processing*—Reduction in the total amount of biosolids to be disposed of is reflected in operational expenditures such as polymer use during dewatering and disposal costs. Processes such as the thermal hydrolysis treatment report significant reductions in the polymer use and increased total solids concentrations of the final dewatered cake.

2.7 Gas Composition

Digester gas is typically composed of approximately 55 to 70% methane and 30 to 45% carbon dioxide (by volume), plus some other minor constituents. However, the methane-to-carbon-dioxide ratio varies depending on the waste composition and the alkalinity and pH of the digester contents. It is important to be able to predict the distribution of the gases that may be produced in the design of systems for gas recovery, cleaning, and use.

2.8 Energy Consumption in Conventional Digesters

The principal energy demands for conventional digesters are fuel for heating, pumped recirculation for heating, and mixing energy.

2.8.1 Digester Heating Requirements

Conventional digesters are typically operated in the mesophilic (32 to 38 °C [90 to 100°F]) temperature range, although thermophilic (50 to 60 °C [122 to 140 °F]) treatment has gained popularity recently. Lower temperatures are seldom used because of the long SRTs required for stable operation in addition to solids destruction and pathogen removal. Accordingly, most conventional facilities are provided with supplemental sludge heaters. Generally, either natural gas, digester gas, or both are used as supplemental fuel for sludge heaters; however, fuel oil is also used. The amount of energy consumed is a function of the incoming waste sludge temperature, digester operating temperature, ambient temperature, and reactor construction and insulation. It should be emphasized that optimizing thickening of sludge prior to digestion is an important energy saving practice as it minimizes the amount of water to be heated.

The quantity of heat required for heating the influent liquid waste stream can be estimated as follows:

$$Q_S = W_S\, C_S\, (T - T_i) \tag{10.6}$$

Where

Q_s = heat required to heat influent waste to the digestion temperature, kJ (Btu/hr);

W_s = mass flow of wet influent sludge (mostly water), kg/h (lb/hr);

C_s = specific heat of influent sludge, kJ/kg · °C (Btu/lb/°F) (typically the same as water, 4.18 kJ/kg · °C or 1.0 Btu/lb/°F);

T = digestion temperature, °C (°F); and

T_i = influent waste temperature, °C (°F) (after heat exchange with effluent, if any).

For example, assuming a typical winter influent temperature for municipal sludge of 50 °F and an operating temperature of 95 °F, Q_s can be estimated as follows:

$$Q_S \text{ (Btu/hr)} = 8.34 \text{ lb/gal} \times q \times 1.0 \times (95 - 50) \qquad (10.7)$$

$$Q_S = 375 \text{ Btu/gal} \times g$$

Where

q = sludge flow, gph.

Note that this calculated amount does not include efficiency losses from transfer of heat from the fuel source (e.g., boiler and heat exchangers) or any benefit from heat exchange with the incoming sludge.

The heat required to compensate for heat losses through the reactor walls can be computed as follows:

$$Q_r = UA(T - T_a) \qquad (10.8)$$

Where

Q_r = heat loss to surroundings, kJ (Btu/hr);
U = composite coefficient of heat transfer, kJ/m$^2 \cdot$ s $\cdot$ °C (Btu/hr/sq ft/°F);
A = area normal (perpendicular) to heat flow, m^2 (sq ft);
T = digestion temperature, °C (°F); and
T_a = ambient temperature, °C (°F).

The composite heat transfer coefficient per unit area, U, can be computed from the transfer coefficient of individual components in series with the direction of heat flow, as follows:

$$U = (1/U_1 + 1/U_2 + ... + 1/U_n)^{-1} \qquad (10.9)$$

Where

U_1, U_2, and U_n = heat transfer coefficients for individual wall components between the sludge within the digester and the air or earth on the outside.

Table 10.3 lists heat transfer coefficients of typical anaerobic digester construction materials. The total heat loss through the digester walls is computed by summing the heat losses through areas of similar construction materials and temperature gradients (e.g, roof or walls exposed to atmosphere, walls exposed to soil, and others).

A typical heat loss value for the Northern United States is 97 kJ/m$^3 \cdot$ h of reactor volume (2.6 Btu/hr per cu ft). Typical heat loss values for middle and southern states are 48 and 37 kJ/m$^3 \cdot$ h (1.3 and 1.0 Btu/hr/cu ft), respectively. With this background,

TABLE 10.3   Heat transfer coefficients for various anaerobic digestion tank materials

Material	Heat-transfer coefficient, U, Btu/hr/sq ft/°F[a]
Fixed steel cover (0.25-in[b] plate)	0.91
Fixed concrete cover (9-in thick)	0.58
Floating cover (wood composition roof)	0.33
Inflatable cover	
Concrete wall (12-in. thick) exposed to air	0.86
Concrete wall (12-in. thick), 1-in. air space, and 4-in. brick	0.27
Concrete wall or floor (12-in. thick) exposed to wet earth (10-ft[c] thick)	0.11
Concrete wall or floor (12-in. thick) exposed to dry earth (10-ft thick)	0.06

[a] Btu/hr/sq ft/°F $\times$ 5.678 = J/m$^2 \cdot$ s $\cdot$ °C
[b] in. $\times$ 25.4 = mm
[c] ft $\times$ 0.3048 = m

Table 10.4 presents typical heating requirements for single-stage sludge digesters in the northern United States for various hydraulic loadings, recognizing that some heat loss can be reduced by additional insulating.

The degree of stabilization is a function of sludge characteristics, digestion temperature, and SRT. In a completely mixed digester (assuming relatively uniform mixing), SRT is equivalent to the hydraulic retention time (HRT) (when the digester

TABLE 10.4   Supplemental heating required for single-stage completely mixed digesters in the Northern United States

Hydraulic retention time, days	Heating of liquid from 50°F to 95°F,[a] Btu/gal[b]	Heat loss from digester, Btu/gal	Total digester heat loss, Btu/gal	Total heat required for heat losses,[c] Btu/gal
10	375	83	458	573
15	375	125	500	625
30	375	250	625	781
50	375	417	792	990

[a] (°F-32)0.555 6 = °C
[b] Btu/gal $\times$ 278.7 = kJ/m^3
[c] Fuel for heating corrected for energy efficiency of boiler at 80%

is operated with decanting, SRT and hydraulic residence time differ). Therefore, by concentrating the feed before digestion, both a substantial energy savings from sludge heating and a higher degree of stabilization from a longer holding time can be realized. In a well-mixed reactor, feed solids concentrations as high as 6% can be handled effectively within the reactor.

A comparison of the effects of feed solids concentration on the heating requirements is presented in Figure 10.2. As can be observed, there is an exponential reduction in feedstock heating requirements with an increased concentration of the feedstock. The effect is further compounded because a thin sludge also reduces SRT (HRT) in the digester during operation as Figure 10.2 also shows. A reduction of SRT (HRT) reduces volatile solids destruction and, consequently, the biogas yield per pound of sludge fed to the digester.

2.8.2   Energy Requirements for Sludge Heaters and Recirculation Pumping

Digester heating is generally provided by an external, boiler-heat exchanger, fueled either by digester gas or natural gas. The efficiency of a sludge heater typically ranges from 70 to 85%. Higher efficiencies are associated with newer equipment that is well

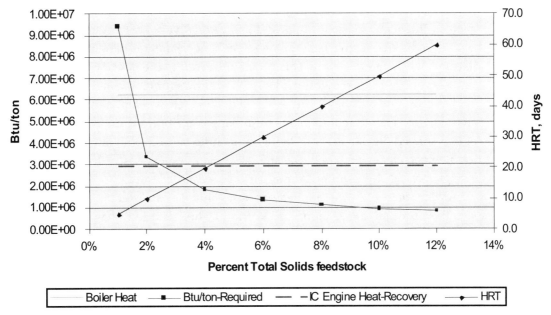

FIGURE 10.2   Effect of thickening performance on heating requirements.

maintained. Heat exchanger scaling decreases efficiency significantly. In addition, electrical energy is required for pumped recirculation and miscellaneous water pumps. Approximately 15.3 to 25.4 mW/W of heater capacity (6 to 10 hp hr/mil. Btu) is required, corresponding to an electrical energy demand of approximately 6 mW/kJ (6 kW/mil. Btu) of heat transferred. Direct steam injection (DSI) provides increased heat transfer efficiency. The City of Crystal Lake, Illinois, recently reported successful operation of a DSI system for their anaerobic digester. No problems of struvite formation or negative impact in the digestion process were observed (Huchel et al., 2006). Digesters at the Back River Plant in Baltimore, Maryland, have used DSI for many years.

2.8.3   Mixing Energy

Digester mixing is accomplished by pumping, or recirculating, the reactor contents. This is typically done using external pumps or compressors to recirculate gas or liquid, or using internal impeller mixers.

With respect to digester mixing, there is great disparity in mixer design and disagreement in perceived effect. In most installations, mixing is relatively ineffective: less than 50% of the total volume is effectively used. Early digester mixer designs were used for scum layer control; thus, they were not intended for blending the entire reactor contents. Deposits of rags and grit in relatively flat-bottom type digesters result in reduced effective tank volume. Frequent cleaning is required to keep many digesters working effectively. This problem has led to the development and recent successes of egg-shaped digesters. This shape allows for more effective recirculation and has proven to enhance the time-VSS-reduction relationship. It can be argued that a reduction in energy requirements for mixing contributes to a lower life cycle cost for egg-shaped digesters. While being more efficient from a mixing and VSS reduction perspective, egg-shaped digesters may be more expensive to build and may have greater heat losses from their larger exposed surface areas.

A wide range of power applied per unit volume of digester has been used in the past. Power for mixing as high as 13 W/m^3 (0.5 hp/1000 cu ft) has been advocated. Values between 5 and 8 W/m^3 (0.2 and 0.3 hp/1000 cu ft) have been most frequently used. More recently, researchers and engineers have taken a closer look at the mixing efficiency in digesters as a result of the industry trend to operate digesters at higher total solids concentrations. Viscosity and shear sensitivity of digested sludge changes significantly with increased solids concentration. Relatively low increments of total solids in the digester sludge, from 2% to 4%, can have significant effects in the mixing behavior of the digester tank. Computational fluid dynamics studies using measured rheological characteristics of the digesting sludge have allowed optimization of mixing of digesters in spite of the increased viscosity of the sludge at these high concentrations. Tracer studies in pilot- and full-scale digesters have confirmed the modeling results. Power use as low as 4 W/m³ (0.14 hp/1000 cu ft) using sequential gas mixing has been reported for digesters operated at 7 to 8% total solids with sludge pretreated with a thermal hydrolysis process.

In general, mixing energy is a relatively minor component of the overall energy expenditure in the digestion process. The principal component in the overall digestion balance is heating of the feedstock.

2.9 Energy Recovery

Digester gas contains 40 to 75% methane, with 60% being a common percentage. Because methane has a higher heating value of 37 000 kJ/m³ (1000 Btu/cu ft), the higher heating value of digester gas is commonly taken to be 22 000 kJ/m³ (600 Btu/cu ft). Approximately 0.75 to 1.25 m³ of digester gas is produced per kilogram of VSS destroyed (11 to 18 cu ft/lb) in anaerobic digestion, with 0.8 m³/kg (13 cu ft/lb) being a common value. The latter is equivalent to a heating value of 17 000 kJ/kg of VSS destroyed (8 000 Btu/lb). Table 10.5 illustrates some common uses of digester gas.

Digester gas is commonly used to provide the heat necessary to maintain digestion temperatures near 35 °C (95 °F) in mesophilic and 55 °C in thermophilic digesters. In areas where there is sufficient excess digester gas, installation of digester gas engines may be warranted. Engines can be used to drive pumps, blowers, or generators. Engine

TABLE 10.5   Common uses of digester gas

Use	Equipment
Digester Heating	Boiler, Heat recovery equipment, heat exchangers
Electric Power Generation	Gas cleaning, IC engine, Microturbine, Turbine, Fuel Cell, Sterling engine, Steam Turbine
Building Heating	Heat recovery equipment, heat exchangers
Air Conditioning	Heat recovery equipment, chiller
Biosolids Drying	Dryer, heat recovery equipment
Biosolids Pasteurization	Boiler, Heat recovery equipment, heat exchangers
Thermal Hydrolysis	Boiler, Heat recovery equipment, direct heat injection
Methane Gas Retail	Gas treatment
Drive Pumps and/or Blowers	Gas engine driven pumps and blowers
Flaring	Flare

cooling water can be used as the source of digester heat when used in conjunction with appropriate heat exchangers. Use of exhaust heat from biogas-driven engines has also been used for biosolids drying or for general facility heating. Steam generated in boilers has been used in biosolids pasteurization processes and thermal hydrolysis of sludge prior to digestion. Biogas is also sold to power stations, industries, or gas utilities. Internal combustion engines, microturbines, and fuels cells have recently become more efficient. Table 10.6 compares the relative combined heat and power efficiencies of the most common technologies.

Biogas usually requires treatment prior to use in some of the equipment presented in Table 10.6. Hydrogen sulfide concentrations vary significantly from system to system depending on the amount of sulfate or other sulfur compounds originally present in the digester feedstock. Siloxanes are silica-based organic compounds that are present in biogas in trace amounts as a result of personal care products present in the sludge. Siloxanes form silica powders during combustion that are deleterious for the performance of some of the equipment and need to be removed from biogas. Biogas is saturated with water vapor from the digester's warm environment. Depending on the downstream use, water vapor might need to be removed as well. A wide range of technologies exist for removal of hydrogen sulfide from biogas. The cost of gas treatment must be considered as part of the operational expenses of the project.

TABLE 10.6   Summary of typical parameters for biogas utilization equipment

| Technology | Net Electrical Efficiency | | Net Thermal Efficiency | | Size Range |
| | Range | Typical | Range | Typical | |
	%	%	%	%	kW
Internal Combustion Engine. ICE	25–45	33	40–49	40	50–5000
Internal Combustion Engine. Lean Burn		37			
Gas Turbines. Combustion Gas Turbines	23–36	30	40–57	40	250–250,000
Microturbines	24–30	27	30–40	35	30–250
Fuel Cells. Phosphoric Acid	36–40	35	30–40	40	200
Fuel Cells. Moleten Carbonate	40–45	50	30–40	40	300–1,200
Steam Turbine	20–30	25	20	45	500–1,300,000
Stirling Engines	25–30	27	45–65	60	1 to 50

TABLE 10.7   Facts for calculating the potential power generated from digester gas

Fact	Unit	Value	Range
Energy content of methane gas	Btu/ ft3	1000	N.A
Methane content of Biogas	%	60	55–65
Unit Biogas generation rate	ft3/PE day	1	0.75–1.25
Per capita wastewater production	gal/PE day	100	70–120
Digester gas yield	ft3 biogas/lb solids destroyed	13	11 to 18

As presented in Table 10.6, conversion of fuel to electricity is relatively inefficient. This is especially true when using a diluted fuel such as digester gas that is contaminated with carbon dioxide and hydrogen sulfide and is saturated with water vapor. Approximately 33% of the heat content of digester gas used in an engine generator goes to producing work, and, at 90% generator conversion, only 30% goes to actual production of electricity. It is also important to note that heat can be recovered from the engine cooling system and the exhaust gas. When heat recovery is incorporated, total energy recovery from biogas is significantly better. Some useful facts for calculating the potential power generated from digester gas are presented in Table 10.7.

3.0   AEROBIC DIGESTION

Aerobic digestion and anaerobic digestion are processes used to stabilize the organic matter in sludge. Unlike anaerobic digestion, which reduces sludge volatile matter through anaerobic biochemical reduction processes, aerobic digestion reduces sludge volatile matter through biochemical oxidation processes using an aerobic environment. Aerobic digestion is energy intensive and produces a sludge that is often difficult and costly to dewater. It also has the disadvantage that it will undergo further anaerobic decomposition, with the production of odor, once oxygen is no longer available. A comparison of advantages and disadvantages of aerobic digestion is presented in Table 10.8.

Wastewater treatment plants that use aerobic sludge digestion rarely use primary clarification because there is no economic benefit. Regardless of whether the primary solids are oxidized in the secondary treatment system or in the aerobic digester, approximately the same amount of oxygen and energy will still be required for oxidation. Primary clarification, then, becomes a needless extra step.

TABLE 10.8   Advantages and disadvantages of aerobic stabilization processes

Advantages	Disadvantages
Moderate degree of stabilization; inactivates pathogens	Often produces heavy surface foam with associated odors
Lessens amount of sludge for final disposal	Sludge not fully stable; will further decompose anaerobically, producing odors
Fairly easy to operate	
Less capital intensive and less maintenance intensive than anaerobic digestion	Requires long solids retention times
	High energy requirements
Side stream generated is not as poor quality as that from anaerobic digestion	Sludge is difficult and costly to dewater

Without primary clarification, however, the volume of an activated sludge aeration tank will have to be increased to accommodate the increased loading. Because aerobic digesters are typically equipped with less efficient diffusers than the secondary process, it is often more desirable to complete as much oxidation of the primary solids as possible in the aeration system rather than in the aerobic digester.

Aerobic digestion of waste secondary sludge is a continuation of the endogenous respiration process that begins during secondary treatment. It requires the addition of oxygen through mechanical aerators or diffusers. Typically, the process requires 1500 to 2000 g of oxygen/kg of VSS destroyed (1.5 to 2.0 lb/lb). The higher value includes oxidation of the ammonia nitrogen released in the respiration process as well as oxidation of the VSS. As proteinaceous material is broken down, nitrogen is released as ammonia and oxidized to nitrate in the aerobic digester. Oxygen transfer efficiencies will depend on the type of aeration device chosen as well as other factors, as discussed in Chapter 8.

3.1   Energy-Saving Opportunities

The following operational strategies exist for minimizing energy expenditure during aerobic digestion of excess sludge:

- Aerobic–anoxic operation,
- Operation at low dissolved oxygen concentration,
- Reduction of digestion time by meeting vector attraction criteria using the specific oxygen uptake rate (SOUR), and
- Assess optimal solids concentration.

In aerobic digestion without anoxic periods, the nitrogenous matter in the digesting sludge is transformed to nitrates according to the following reaction:

$$C_5H_7NO_2 + 7\,O_2 + HCO_3^- = 6\,CO_2 + 4H_2O + NO_3^- \qquad (10.10)$$

In aerobic–anoxic operation, the aeration to the digester is turned on and off alternatively. During the anoxic periods, denitrification takes place, converting the nitrates that accumulated during the aerobic period into nitrogen gas. The overall digestion process during aerobic–anoxic operation can be represented by

$$C_5H_7NO_2 + 5.75\,O_2 + HCO_3^- = 6\,CO_2 + 4H_2O + N_2 \qquad (10.11)$$

A 20% reduction in oxygen is obtained, from 2000 g of oxygen/kg of VSS destroyed (2 lb/lb) to 1600 g of oxygen/kg of VSS destroyed (1.6 lb/lb). In addition, there is a recovery of alkalinity by removing nitrates that acidify the digester contents. There is no need to mix during the anoxic period, further reducing energy use if a high enough concentration of MLSS is maintained in the digester.

Operation at low dissolved concentration (0.1 to 0.5 mg/L) induces simultaneous nitrification–denitrification, obtaining a similar effect as presented previously with no need for alternating aeration cycles. A further advantage of this operational modification is reducing the aeration expenditure by running at a lower dissolved oxygen concentration. A disadvantage of low dissolved oxygen operation, however, is the reduction in the VSS destruction rate that has been reported. In order to overcome this effect, the digesters can be run at higher mixed liquor concentrations, effectively increasing HRT in the system.

Additional energy-saving opportunities are the application of dissolved oxygen controls similar to those described in Chapters 8 and 9, and the application of air proportional to the digester level instead of a constant rate.

Another modification of operation that is available is based on Option 4 of U.S. EPA's 40 CFR Part 503 regulations for compliance with vector attraction reduction. Vector attraction reduction can be achieved by obtaining a SOUR of 1.5 mg oxygen/g of TSS/hour. In some cases, as when WAS comes from a plant with long SRTs (e.g., extended air) or a BNR plant, the sludge to be digested has a low SOUR to begin with. In those cases, the SOUR criteria are often achieved before the 38% reduction criteria. The SRT in the digester can be reduced accordingly, generating savings in the energy expenditure. The Class B requirements for pathogen density might then control the overall retention time in digesters operated according to the SOUR requirements.

Lastly, it is important to note that as the solids concentration increases in an aerobic digester, the alpha factor (the ratio of the oxygen transfer coefficients in wastewater versus clean water; see Chapter 8) decreases. A high solids concentration can result in significantly decreased oxygen transfer efficiencies and increased power requirements for the aeration air. Often, coarse-bubble diffusion is used because virtually no gain in net power requirements per unit volume of air added is achieved due to the low alpha factor (Schoenenberger et. al., 2003). The plant designer and operator should consider the impacts on air requirements and energy use with increasing sludge solids concentrations.

3.2 Autothermal Aerobic Digestion

Autothermal thermophilic aerobic digestion (ATAD) is a modification of the conventional aerobic digestion process in which high concentrations of feed solids are used in an insulated reactor. Aerobic digestion, like any other combustion process, is an exothermal process that releases energy in the form of heat. Heat is also generated in conventional digestion, but it is quickly dissipated in the higher volume of water associated with the sludge as a minor, almost imperceptible, temperature rise.

In ATAD, the reactor volume is smaller, the feed solids concentration higher, the retention time lower, and the reactor enclosed and insulated. For municipal sludges, a pre-thickener is necessary. A minimum total volatile solids of 3 to 4% and a total solids content of less than 7% are typical requirements. The result is a rising temperature that accelerates the digestion reaction. In colder climates, reactors may have to be provided with supplemental heat to maintain the desired temperature. Hydraulic retention times are generally on the order of 8 days, but may be as high as 15 days. The reactors must be arranged in a series of at least two reactors. Energy requirements for mixing are one order of magnitude higher than is typically used for anaerobic digesters. Kelly and Warren (1995) report values as high as 130 W/m^3 (5 hp/1000 cu ft) in contrast to values of 7 W/m^3 (0.25 hp/1000 cu ft) for anaerobic digesters.

Autothermal thermophilic aerobic digestion has been shown to meet U.S. EPA's 40 CFR Part 503 regulations for Processes to Further Reduce Pathogens. A combination of ATAD and anaerobic mesophilic digestion has been implemented for obtaining Class A product. In this application, a short aerobic residence time of 18 to 24 hours is used. Heat exchanger is used to preheat the raw sludge to the ATAD system.

Kelly and Warren (1995) reported that ATAD reactors are not truly aerobic and produce off-gases containing short chain volatile fatty acids, dimethyl sulfide, dimethyl disulfide (the same type of odorous byproducts given off by compost systems), and hydrogen sulfide. This may require that off-gases from ATAD processes be scrubbed to remove odors, adding energy requirements and costs to the system input. Volatile fatty acids in ATAD have been proposed as a resource to be used in biological phosphorous removal processes. The dewaterability of ATAD processed sludges is susceptible to the availability of primary clarifiers in the flow stream. Plants have converted from conventional digestion to ATAD and have found that their existing dewatering units' belt presses have required tripling the polymer dosage and obtained considerably "wetter" cake. This phenomenon has been confirmed through correspondence with a major belt press supplier.

4.0 INCINERATION

Incineration is a process in which dewatered sludge is dried and burned to reduce the residual matter to inert ash having a greatly reduced volume. The process energy requirements vary significantly, depending on the type of sludge being processed and its moisture and organic content as it is fed into the incinerator, the type of incineration technology used, and the air pollution requirements and operating practices. Evaporating water is energy intensive. In most instances, supplemental fuel is required to sustain combustion, but in all cases large quantities of heat are involved with potential for energy recovery.

The most common types of incinerators are the multiple-hearth furnace (MHF) and the fluidized-bed furnace (FBF). Generalized schematics for each of these incinerator types are shown in Figures 10.3 and 10.4. Nearly all new thermal oxidation

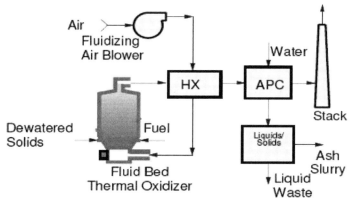

FIGURE 10.3 Multiple hearth process schematic.

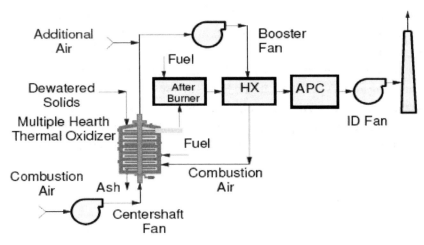

FIGURE 10.4   Fluid bed process schematic.

installations in the last 20 years have used fluidized bed technology. From an energy perspective, both types of incinerators exhibit many similarities. For example, they start with the same type of wet sludge and end with the same type of ash, involve combustion of the sludge volatile matter with oxygen from air, involve the same type of thermodynamic principles, and have similar modes of heat loss. Operation of the MHF is somewhat more complex than that of the FBF in that sludge drying, combustion, and ash cooling take place at various levels in the incinerator as the material passes from top to bottom, while drying and combustion occur in the same single chamber of the FBF. Because drying occurs in the upper hearth of the MHF, the exhaust gas from the incinerator (without an afterburner) should have a temperature in the range of 500 to 650 °C (900 to 1200 °F), which is significantly cooler than the temperature of the exhaust from the FBF that is closer to combustion temperatures at 750 to 850 °C (1400 to 1600 °F). Consequently, to sustain this higher temperature, more heat is sometimes required for combustion in the FBF. However, the FBF will have fewer pollutants in the exhaust, and the potential for energy recovery from the exhaust gas can be greater. An additional consideration is the higher requirements for excess air levels in MHFs than in most FBFs, which can take advantage of high combustion air preheat levels and be designed to operate at near-autogenous conditions.

4.1 Feasibility of Incineration

In many areas, the scarcity of available land for land application of biosolids or the availability of a landfill for disposal drives the evaluation of alternative routes for biosolids processing. Incineration reduces the volume and weight of wet sludge cake by more than 90%, thus reducing the area required for residuals disposal. Recently, advances have been made in the ability to recover heat from the incineration process and reuse the energy throughout the WWTP. This has increased the cost-effectiveness of the incineration process. Recent improvements in thickening and dewatering technology have made it possible to obtain a drier cake from primary waste activated sludge mixtures. The drier cake reduces the auxiliary fuel requirements of the incineration process, making incineration less costly. Improvements in emissions-control technology have made it possible to meet existing emissions standards consistently. However, some states are regulating additional contaminants that have increased the capital and operating costs of compliance with air permits (Burrowes and Bauer, 2004).

4.2 Air Emissions

Incinerators in the United States are required to comply with the New Source Performance Standards, National Ambient Air Quality Standards, National Emissions Standards for Hazardous Air Pollutants, and 40 CFR Part 503 Standards for the Use and Disposal of Sewage Sludge. Air permits issued by federal, state, or local authorities include conditions that are negotiated and that satisfy the foregoing standards. Most incinerators have satisfied their air permits through the use of wet venturi/tray or packed bed scrubbers and, in some cases, with wet electrostatic precipitators and afterburners or regenerative thermal oxidizers. These air pollution control devices or combinations thereof have satisfied best available control technology (BACT) assessments that have targeted particulate matter, sulfur dioxide, and heavy metals. However, some states are requiring BACT for mercury, volatile organic compounds (VOCs), and NOx. This may require the use of other air pollution control technologies (Burrowes and Bauer, 2004).

A factor that has greatly reduced emissions from incinerators is the industrial pretreatment program. The high concentration of pollutants that could contaminate wastewater solids have been controlled at the source and eliminated from sewers. This has translated to a reduction in the potential for emissions at incineration facilities.

4.3 Process Stability

Stability and control are critical to the efficient operation of sludge combustion processes. One cause of process instability is an unsteady feed rate to the furnace, which results from variations in the performance of dewatering facilities. This condition is exacerbated by the discharge of sludge cake from the dewatering system directly to the furnace feed conveyor, with no intermediate surge storage, such as occurs at MHF installations (Lewis et al., 1989).

Because of possible blowback from the pressurized combustion chamber, fluidized bed installations have always had some degree of surge storage or *system capacitance,* such as a hopper above the sludge feed screw or pump. Providing a similar, though small level of sludge feed surge storage (30 to 45 minutes) can improve the overall operation of an MHF installation (Lewis et al., 1989).

The sludge inventory of an MHF is large, typically approaching 1 hour, whereas that of an FBF can be measured in seconds. Each hearth in an MHF has its own sludge inventory and is affected differently by process upsets. A greater or smaller sludge inventory within a furnace will not necessarily lead to improved process stability. However, changing the inventory of sludge over a relatively short period of time does affect process stability (Lewis et al., 1989).

The 40 CFR Part 503 regulations, as enacted for sludge combustion, primarily affect emissions of various metals and VOCs and, further, require specific devices for monitoring key system operating parameters such as feed rate, exhaust oxygen, combustion process temperatures, and total hydrocarbons. In reality, performance of the combustion and scrubbing systems cannot be optimized independently because the performance of both systems is unalterably linked to the stability of the overall process. Enhancing process stability must be considered a fundamental element in meeting stringent air pollution control regulations.

4.3.1   Multiple-Hearth Furnace

The multiple-hearth furnace is the most common furnace in use for incinerating sludge. It involves feeding dewatered sludge at the top of the incinerator and allowing it to pass through the three stages of drying, combustion of organic matter, and cooling of the ash as it passes from the top hearth to the bottom hearth. The MHF is susceptible to upset and requires vigilant operational control. When either feed rate or moisture content changes abruptly, the steady-state conditions throughout the incinerator can quickly be thrown into imbalance.

Because of the irregular route of airflow through the incinerator, short-circuiting and poor air mixing can occur throughout the unit, requiring excesses of air for sludge combustion. This inefficient use of air results in a large heat loss for MHFs.

In February 1991, U.S. EPA published 40 CFR Part 503, which established rules governing standards for the disposal of wastewater sludge by land application, land disposal, and incineration (U.S. EPA, 2009). These regulations require that exhaust gas oxygen analyzers be used with all MHF (and FBF) systems, allowing the integration of oxygen control into furnace operation. The 40 CFR Part 503 regulations also require that at least two independent thermocouples be provided in the combustion zone of MHFs. Because the combustion zone is dynamic, this will affect several hearths in the furnace. Even using two thermocouples may fail to provide reliable and representative temperature measurements unless the thermocouples are properly placed in the furnace and are of the appropriate insertion lengths (Sieger and Maroney, 1977).

Many agencies with multiple hearths have required retrofits and operational changes to comply with more stringent air emissions. These modifications have included afterburner chambers, exhaust gas recirculation, add-on thermal oxidizers, and improved scrubbers. Other facilities have increased afterburner operating temperatures to reduce "yellow plume" emissions or cyanide concentrations in the scrubber water. Both of these emissions are signs of incomplete combustion.

4.3.2 Fluidized-Bed Furnace

For FBFs, startup fuel requirements are low, and little fuel is required for startup following overnight shutdowns. In an FBF, the sand bed acts as a large heat reservoir, minimizing the amount of fuel required to reheat the system following shutdown, which makes FBFs good options for intermittent operation. Exhaust temperatures of FBFs exceed 750 °C (1400 °F), so an afterburner (which requires supplemental fuel) to comply with air pollution regulations is often not required. Problems with FBFs include the feed system and temperature control with high-energy feeds such as dewatered scum (Lewis et al., 1989).

Violent mixing in the fluidized bed ensures the rapid and uniform distribution of fuel and air and, consequently, good heat transfer and combustion. The bed itself provides substantial heat capacity, which helps to reduce short-term temperature fluctuations that may result from varying feed rates and feed heating values. Sludge particles remain in the sand bed until they are reduced to mineral ash. The violent motion of the sand in the bed grinds the ash material until it is so fine that it is readily stripped from the bed by the upflowing gases (Sieger and Maroney, 1977).

Even though FBF systems are typically more stable than MHF systems, process stability and control are linked to providing a steady rate of feed. In addition, the nature of the bed zone, because of the large heat sink capacity of the bed sand material, makes the development of an appropriate operating and control philosophy more complex. Bed temperatures tend to move deceptively slowly, often leading the operators to develop a false sense of security relative to normal operating routines and further complicating the configuration of control systems.

Another important difference for the fluid bed is the ability to better control combustion air. Fluid bed systems typically need only 30 to 50% excess air as compared to 50 to 150% for MHF systems.

A key consideration in developing a reliable control philosophy for FBF systems centers on providing representative measurements of bed temperature. The 40 CFR Part 503 regulations require only a single thermocouple to be located in the bed zone; however, this is inadequate in most systems. While FBF systems provide excellent vertical mixing because of the turbulence of the sand bed, lateral mixing is relatively poor in most cases, making bed feed distribution an important design parameter. Even with good distribution, multiple thermocouples should be used at strategic elevations and locations around the perimeter of the bed (Lewis et al., 1989).

4.4 Heat Requirements

Regardless of the type of incinerator used, heat balances are used to estimate energy use and requirements. The major heat inputs and losses that must be considered for incinerators are given herein. Heat gains (inputs) result from

- Combustion of sludge volatile matter,
- Combustion of auxiliary fuel, and
- Preheated combustion air (recirculated cooling air or heat exchange from exhaust).

Heat losses (outputs) include losses resulting from

- Water in sludge vaporized to exhaust,
- Exhaust of hot combustion gases,
- Exhaust of heated excess air and nitrogen associated with combustion air,
- Radiant heat from incinerator shell,
- Heated ash removed from the furnace,
- Vented shaft cooling air (MHF), and
- Gas cooling spray water vaporized to exhaust (FBF).

The major factors affecting auxiliary fuel use are

- Water in feed sludge cake,

- Volatile solids content of the feed,

- Incinerator exhaust temperature,

- Excess air for sludge combustion, and

- Shaft cooling air (MHF only).

Energy losses associated with water are high and, therefore, significant regarding incinerator fuel use and energy recovery. Although water is formed as a product of combustion and contained as humidity in combustion air, most water in the incinerator heat balance is from free water contained in the sludge.

4.4.1 Heat Losses Associated with Water

Because heat losses caused by water are so important to energy use and recovery, one should understand why these losses can be so large and energy recovery so poor. Even in the best dewatering applications, resulting sludge cake contains more water than solids. In wastewater applications, rarely will sludge feed cake in an incinerator have less than 60% water, and frequently it will contain more than 75% water.

As the sludge is heated from ambient temperatures to exhaust temperatures, the heat capacity of its contained water undergoes three significant changes as the water is converted from liquid to gas (heat capacity represents the amount of heat required either to raise the temperature of a unit weight of the material or to transform a phase change). Water in the sludge remains as a liquid until its temperature is increased to the boiling point, at or near 100 °C (212 °F). At this point at standard atmospheric pressure, a large amount of heat is required to evaporate the water while the temperature of the liquid remains near 100 °C (212 °F). Once evaporated, the temperature of the vaporized water can again increase. To determine the total heat required for heating water from ambient to exhaust temperatures, one must use steam tables or perform three separate heat calculations as follows:

- *Water as liquid*—heating of the water (moisture) in the raw sludge from ambient to the boiling point of water. Approximately 4.18 kJ (1 Btu) is required to raise the temperature of 1 kg (1 lb) of the water in the raw sludge 1 °C (1 °F). This 4.18 kJ/kg·°C (1 Btu/lb/°F) is the heat capacity of water.

- *Water boiling*—at the boiling point, water is converted from a liquid to a gas (steam) without a temperature change, but requires a large input of heat energy. This phase change requires that 2257 kJ/kg of water (970.3 Btu/lb) be supplied to convert the water from liquid to gas. (This 2257 kJ/kg [970.3 Btu/lb] required for the phase change of water from liquid to gas is called the *latent heat of vaporization*.)

- *Water as gas*—once the water is converted to steam, more heat is required to raise the temperature of each kilogram (pound) of steam 1 °C (1 °F) to the exhaust temperature. The heat capacity of steam (water as a gas) increases as its temperature rises, but is on the order of 1.97 kJ/kg·°C (0.47 Btu/lb/°F), which is less than half the heat capacity of water as a liquid.

Because the steam created is at atmospheric pressure, it cannot do useful work like the pressurized steam in a steam engine. Actually, because the exhaust temperature of the flue gas is rarely reduced to less than 200 °C (400 °F) with heat recovery devices, most of the heat energy used for water evaporation is wasted in the exhaust gas in the form of uncondensed steam. While it is technically possible to recover this heat, it is impractical for several reasons:

- Large heat exchanger surface areas would be required for condensation of the water vapor;

- To extract most of the heat, the temperature of any heat transfer medium would be below 100 °C (212 °F) and would be of marginal value; and

- The corrosive nature of the exhaust gases in a condensing heat exchanger would require costly materials of construction.

4.4.2   Wastewater Solids Energy

The organic content of dewatered wastewater solids can range from 50 to 85% depending on the upstream treatment processes, such as the mixture of primary to waste activated sludge and the extent of the digestion process. In addition, the energy content of the organic material can vary from 19 800 to 30 200 kJ/kg (8500 to 13 000 Btu/lb) of volatile solids; however, for sludge produced in municipal WWTPs, a typical range is 23 300 to 25 600 kJ/kg (10 000 to 11 000 Btu/lb) volatile solids. Raw solids usually have higher energy content than digested solids due to the higher content of volatile solids in the former. The importance of having drier solids in order to achieve autogenous (without supplemental fuel) combustion is illustrated in Figure 10.5.

4.4.3   Autogenous Combustion

An example calculation of typical fluidized-bed incinerator energy use as a function of varying total solids contents is presented in Figure 10.5. Typical heat content of a

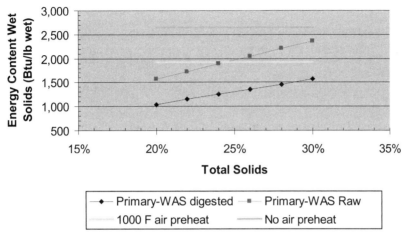

FIGURE 10.5   Effect of total solids on autogenous combustion.

wet sludge for autogenous combustion in fluidized-bed systems is 4478 kJ/kg (1925 Btu/lb) for systems with preheat combustion air to 540 °C (1000 °F) and 6164 kJ/kg (2650 Btu/lb) for wet solids with systems with no air preheat. The figure suggest that for a feedstock with 75% volatile solids (a mixture of raw primary and WAS), autogenous is achievable around 26% solids. A well-digested sludge with a volatile solids content of 50% would require a higher air preheat to achieve autogenous combustion. One of the inherent advantages of fluid beds is that they can operate with high-temperature (540 to 650 °C [1000 to 1200 °F]) combustion air.

4.4.4   Electrical Energy Use

Electrical energy is required to drive motors of fans, pumps, conveyors and other drives as well as to energize the air pollution control equipment. Electrical energy requirements are most influenced by the fan power requirements, which, in turn, are a function of the quantity of drying air required and how much exhaust air is treated by the air pollution control equipment. Pressure to move combustion air and exhaust gases through the system is directly affected by the air pollution control equipment. Installed electrical power in the range of 160 to 230 kW/Mg (200 to 275 hp/dry ton) has been reported for FBF using air preheaters and venture/tray or packed bed scrubbers (Burrowes and Bauer, 2004). Electric energy use ranges from 790 to 1440 kJ/kg (220 to 400 kWh/dry tonne) for FBF and from 470 to 860 kJ/kg (130 to 240 kWh/dry tonne) for MHF.

4.4.5 Energy Recovery

In a FBF with a heat exchanger for preheat, exhaust gases leave the system at a temperature on the order of 540 °C (1000 °F). Recovery of energy from heat exchange between 540 °C (1000 °F) and 200 °C (400 °F) can be done by including a steam boiler and a turbine/generator. Another alternative is to use a thermal oil heating heat exchanger; the high temperature of the oil can be used for sludge drying, building/process heating, preheating feed to centrifuges (higher cake solids and lower polymer use), or for an organic Rankine cycle power generation system. The latter one is similar to a steam turbine, but has certain process and operating advantages. The power recovery ranges from 1200 to 2400 kJ/kg (300 to 600 kWh/dry ton). This variation is due to the extent of preheat necessary. As discussed previously, it is usually not practical to cool the exhaust gases below 200 °C (400 °F) for energy recovery.

5.0 DRYERS

Thermal drying technology is based on the removal of water from dewatered biosolids, which accomplishes both volume and weight reduction. The added benefit of thermal drying is that it typically results in a product with a significant nutrient value. Typically, dewatered biosolids (at approximately 18 to 30% dry solids content) are delivered to a thermal drying system, where most of the water is removed via evaporation, resulting in a product containing approximately 90% solids. In the thermal drying system, the temperature of the wet solids mass is raised so that the water is driven off as a vapor. By removing most of the water from the solids, thermal drying results in a significant reduction of both volume and mass.

Significant thermal energy must be transferred to the solids to increase temperature in the drying process. This energy can be provided by the combustion of a variety of fuels (natural gas, digester gas, heating oil, wood, and so on), by a reuse of waste heat, or by conversion of electrical power into thermal energy.

The high temperatures used in thermal drying ensure that the U.S. EPA time and temperature requirements for pathogen kill are met. Drying also meets the U.S. EPA vector attraction reduction standards by desiccating the wastewater solids to greater than 90% solids (or to greater than 75% solids if the solids have been previously stabilized). Although high temperatures are used in thermal drying, the temperatures are generally low enough to prevent oxidation (burning) of the organic matter. Thus, most of the organic matter is preserved in the dried material.

Sludge can be dried in mechanical dryers to

- Produce a dry pulverized material for use as a soil conditioner or fertilizer,
- Remove water before incineration, pyrolysis or gasification,
- Reduce the total volume and weight of the sludge before reuse/disposal, or
- Produce a fuel that can be used as coal replacement in cement kilns or power plants.

Dewatered sludge cake should be used as the dryer feed. Final cake moisture from the dryer depends on the intended use of the sludge following drying. *Scalping* is a term used when cake solids are only increased an additional 5 to 10%. This makes sense where waste heat from an incinerator, for example, can be used to provide evaporation of some of the cake moisture so that the feed sludge burns autogenously in the incinerator. For landfill disposal, intermediate dewatering to 50 to 65% cake solids may be appropriate. Without pelletizing, higher cake solids can result in severe dusting and a light, fluffy product, which, among other problems, becomes difficult to handle. Some of the commercially available dryers are specially designed to produce a granular or pelletized product that is low in dust.

Sludge enters a glue or sticky phase when it gets above 30 to 40%, depending on the type of sludge (in some instances, the glue phase has been observed in the 40 to 60% solids range). To prevent the sludge from sticking to the dryer parts, some of the finished material must be recycled back and mixed with the feed sludge to make the feed material more friable.

5.1 Purpose

Dryers have been used to produce a dry material containing organic matter and nutrients that can be used as low-grade fertilizer and soil conditioner or supplemented with nutrients to produce a commercial-quality fertilizer. Generally, the sludge used in the dryer is biologically unstabilized and, once wetted, becomes biologically active, produces heat, and often produces odors. Nevertheless, a large amount of this dried wastewater sludge originating from large U.S. cities has been used for agricultural purposes for many years. An alternative use for the dried sludge is to substitute fuel for industrial or power plant uses.

Dryers do not necessarily have to remove all the water in the sludge and are often used to remove only a portion of the water. This can be done by partial drying of the entire sludge stream or by combining wet sludge with dried sludge to attain the desired product. Both the costs of landfill disposal and incineration can be reduced by partial drying of sludge.

Where landfill space is expensive, as in the Northeastern United States, it is economically viable to dry sludge before disposing of it in a landfill. Because it takes approximately 75 L (20 gal) of fuel oil to evaporate 0.9 Mg (1 ton) of water, drying can be an attractive option even in the absence of waste heat.

Sludge that is dried to a high degree is often voluminous, dusty, and difficult to handle. For landfill disposal, it is generally more feasible to dry the sludge to solids of 80% or less for ease in handling.

Because drying is part of the incineration process, it is, in most cases, more efficient and cost-effective to design an incinerator to do both tasks rather than using a combination dryer/incinerator. Exceptions to this are

- Existing installations that are currently limited in incineration capacity because of high water content of sludge. In this instance, it may be found that adding a dryer to reduce a portion of the sludge water can maximize production of capital already in place. Production of the incinerator can be increased by predrying all or a portion of the sludge.

- Incinerator installations that have afterburners with heat recovery and no useful purpose for the recovered heat. Heat that would otherwise go to waste could be used to dry the sludge to make the overall process more efficient and more economical.

- Anaerobic digestion operations in which surplus gas can be used to provide heat for drying.

- Co-location of heat dryers along existing power plants or cement plants with available waste heat.

In general, separate drying of sludge is inefficient and not cost-effective when used in conjunction with incineration. However, there are specific applications for which this is not true. These applications are typically the result of unusual local or economic conditions or limitations of existing expensive incineration equipment.

5.1.1   Energy Use

Heat balance calculations for dryers are similar to those for incineration presented earlier in this chapter. Because there is no heat generated in the process, most of the heat loss is attributable to the amount of water in the sludge to be dried. Recovery of heat from the dryer flue gas is often impractical and, therefore, is seldom attempted. Dryers can be coupled to other thermal processes such as a genset used for electricity generation from digester gas or to an incinerator to recover some of the heat produced there.

Energy consumed in a thermal drying system typically includes fuel/thermal energy and electric power to operate the equipment. The thermal energy consumption is based on the amount of water to be evaporated and the thermal efficiency of the drying system. Thermal efficiency of drying systems may range from approximately 3300 kJ/kg (1400 Btu/lb) of water evaporated to 4000 kJ/kg (1700 Btu/lb) of water evaporated. Cost of fuel is one of the largest costs for any thermal drying system. If digester gas or a waste heat source is available, considerable savings in fuel costs can be realized. Similarly, significant reduction in energy use for drying is realized from having a dryer cake produced in the dewatering step. Figure 10.6 illustrates the effect of a drier feedstock to the dryer.

Dryers have significant heat and electrical requirements. In addition to a heat source, dryers require pumps or blowers for circulating the heat transfer medium as well as drives, mixers, conveyors, and other auxiliary devices. The amount of power consumed by thermal drying systems with identical capacity may vary depending on the type of thermal drying system. (i.e., direct-type dryers are likely to consume more power than indirect dryers). Direct drying systems typically have a connected load of approximately 0.01 to 0.5 kJ/kg (10 to 65 kWh/1000 lb) of water evaporation capacity. The lower load is for a tray dryer and the higher load for a fluidized-bed dryer. This electric load is equivalent to an increase in energy requirement between 2.5 and 16% of the heat demand for water evaporation.

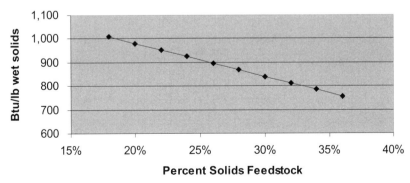

FIGURE 10.6   Thermal energy demand for drying to 90% solids.

6.0 REFERENCES

Burrowes, P.; Bauer, T. (2004) Energy Considerations with Thermal Processing of Biosolids. *Proceedings of the Bioenergy Workshop—Permitting, Safety, Plant Operations, Unit Process Optimization, Energy Recovery and Product Development;* Cincinnati, Ohio, Aug 11–12; Water Environment Federation: Alexandria, Virginia.

Huchel, J.; Van Dixhorn, L.; Podwell, T. (2006) Anaerobic Digesters Heated by Direct Steam Injection: Experience and Lessons Learned. *Proceedings of the 79th Annual Water Environment Federation Technical Exhibition and Conference;* Dallas, Texas, Oct 18–22; Water Environment Federation: Alexandria, Virginia; pp 407–414.

Kelly, H.; Warren, R. (1995) What's in a Name?—Flexibility. *Water Environ. Technol,* **7** (7), 46.

Krugel, S., Parrella, A., Ellquist, K., Hamel, K. (2006) Five Years of Successful Operation: A Report on North America's First New Temperature Phased Anaerobic Digestion System at the Western Lake Superior Sanitary District (WLSSD). *Proceedings of the 79th Annual Water Environment Federation Technical Exhibition and Conference;* Dallas, Texas, Oct 18–22; Water Environment Federation: Alexandria, Virginia; pp 357–373.

Lewis, F.M.; Lundberg, L.A.; Haug, R.T. (1989) Design, Upgrading and Operation of Multiple Hearth and Fluidized Bed Incinerators to Meet the EPA 503 and Other Proposed New Regulations. Paper presented at Sludge Composting, Incineration and Land Application: Burning Issues and Down-to-Earth Answers; Virginia Water Pollution Control Association: Richmond, Virginia.

Schoenenberger, M.; Shaw, J.; Redmon, D. (2003) Digester Aeration Design at High Solids Concentrations. Paper presented at the 37th Annual Wisconsin Wastewater Operators Association Conference, Wisconsin Dells, Wisconsin.

Sieger, R. B.; Maroney, P. M. (1977) Incineration-Pyrolysis of Wastewater Treatment Plant Sludges; U.S. Environmental Protection Agency *Design Seminar for Sludge Treatment and Disposal;* U.S.Environmental Protection Agency: Washington, D.C.

U.S. Environmental Protection Agency (2009) *Code of Federal Regulations;* 40 CFR Part 503; U.S. Environmental Protection Agency: Washington, D.C.

U.S. Environmental Protection Agency (1979) *Process Design Manual, Sludge Treatment and Disposal;* U.S. Environmental Protection Agency, Municipal Environmental Research Laboratory: Cincinnati, Ohio.

Walsh, M. J.; Pincince, A. B.; Niessen, W. R. (1990) Energy-Efficient Municipal Sludge Incineration. Water Environ. Technol., 2 (10), 36.

Willis, J.; Schafer, P.; Switzenbaum, M. (2005) The State of the Practice of Class-A Anaerobic Digestion: Update for 2005. *Proceedings of 78th Annual Water Environment Federation Technical Exhibition and Conference;* Washington, D.C., Oct 29–Nov. 2; Water Environment Federation: Alexandria, Virginia; pp 886–903.

Willis, J.; Schafer, P. (2006) Advances in Thermophilic Anaerobic Digestion. *Proceedings of the 79th Annual Water Environment Federation Technical Exhibition and Conference;* Dallas, Texas, Oct 18–22; Water Environment Federation: Alexandria, Virginia; pp 5378–5392.

Wong, V.; Bagley, D. M.; MacLean, H. L.; Monteith, H. (2005) WERF: Comparison of Full-Scale Biogas Energy Recovery Alternatives. *Proceedings of the 78th Annual Water Environment Federation Technical Exhibition and Conference;* Washington, D.C, Oct 29–Nov. 2; Water Environment Federation: Alexandria, Virginia; pp 6480–6494.

Chapter 11

Energy Management

(continued)

1.0 ENERGY MANAGEMENT PLAN OVERVIEW

1.1 Energy Management Plan

Electricity, or energy in general, represents either the largest or second-largest expense in a water and/or wastewater utility (W/WWU) budget alongside personnel costs. Many W/WWU managers and supervisors have little or no control over personnel costs, but typically have control over energy consumed at their facility responsive to equipment and structures at the facility. Therefore, cost control through energy management should be a high priority for all W/WWU managers.

Typically, the first approach taken when it is decided to reduce energy use is to make a casual W/WWU energy survey to determine if and where there are excesses and abuses. Lights are turned off, thermostats are adjusted, minor changes are made in operation, and a more thorough look at W/WWU operations may be initiated. Often, process changes are identified, assessed, and implemented that result in reduced secondary treatment energy consumption, reduced solids management energy consumption, reduced pumping energy consumption, and other reductions. However, without continuous energy awareness and management, all successes achieved in these efforts disappear over time. This is because the operation typically drifts back to what has been considered normal, safe, or easy. Also, in many locations, energy is not considered an important component of a W/WWU's operation or equipment selection process. Therefore, industry continues to develop W/WWU projects that do not include energy efficiency in the design process.

Many suggestions for reducing energy consumption at W/WWUs have been published and many of these suggestions can be put into practice. However, for continuing success, there must be an energy management program.

Energy management is not as complicated as it might seem. Once it has been determined that energy management is necessary, a program can be set up using simple steps such as the following:

- Become aware of your energy consumption,
- Provide education in energy efficiency,
- Gather data,
- Analyze data,
- Learn what information the data provides,
- Create a plan,
- Implement the plan, and
- Establish a timetable to continually revisit and update the developed plan.

The developed plan should never be considered complete or finished because efficiency opportunities are always available.

An energy management program consists of four basic steps. The first two steps pertain to the energy survey/audit. They define what exists, current operating practices, and where and how energy is being used. This information is needed for further decisions. The third step sets the direction for the program and determines whether additional expenditures are required for new or replacement meters or equipment or for professional assistance. The last step involves implementing the plan. The following points describe in more detail the steps for an electrical energy management program.

There are several definitions or models that are available to present a comprehensive and effective energy management plan/program. The following are characteristics that should be included in a plan:

- Strong management support (leadership and allocated resources);
- Integrated energy management process acceptance (council floor to operations personnel);
- Process sub-metering of energy use and budget allocation of energy use;
- Awareness, consideration, and inclusion of energy in process and facility design; and
- Fine-tuning of operations to control energy demand and consumption.

An energy management plan should have the following characteristics customized to a facility:

- *Energy policy*—Create a policy to demonstrate commitment to energy efficiency;

- *Goals and targets*—Set quantifiable goals;

- *Project management*—Present inclusion of energy efficiency in responsibilities of management;

- *Facility profile*—Monitor process and energy use (see Section 2.4.4, "Data Trending," later in this chapter);

- *Equipment and process profile*—Monitor energy use by equipment and process system;

- *Best practices*—Obtain and adopt energy best practices for all activities on the facility; and

- *Continual improvement*—Continually revisit the developed plan to remain proactive and not fall into a reactive mode.

These tools were developed and implemented through an industrial energy efficiency program, Wisconsin Focus on Energy's Industrial Program (Madison, Wisconsin), and presented as practical energy management.

1.2 Energy Awareness—Understanding Energy Use

Treatment facility personnel need to develop a basic awareness of how energy is consumed at their facility. Treatment personnel know electricity, how it works, and the associated safety procedures that must be followed to work with it. Although most states require knowledge of electricity to obtain an operator's license, what is not required is a knowledge or awareness of how energy is used. Knowledge of how energy is used at a facility will provide treatment personnel with a valuable tool to address budget control, thereby helping their facility become a community leader as an environmental steward. The facility that meets all of its necessary effluent limits, while at the same time doing it with minimal energy consumption, can be in the forefront to challenge other facilities to improve their energy efficiency. In addition, by becoming aware of energy consumption, a facility can be a leader in providing guidance to industries in the community, showing them the value gained through energy efficiency.

It is necessary for treatment facility staff to learn the information that is provided to them on the monthly energy bill, beyond the "amount due" line. Operations as well as administration and management staff need to become familiar and knowledgeable with all information provided on the bill. Items such as "kilowatt-hours [kWh], kilowatts [kW], on-peak, and off-peak are just a few of the items that treatment facility personnel need to be knowledgeable of to become aware of energy consumption. (Details on understanding your energy bill are presented in Section 2.1.1 in this chapter and in Chapter 2).

1.3   Available Energy Computer Modeling

This information is presently under review and identification as a component of the Water Environment Research Federation (WERF) Operations of Wastewater and Solids Operations Challenge. Presently, there are computer-modeling programs available for developing and comparing design concepts. Integrating energy consumption and its related costs are just beginning, and their availability will be an item to be identified through the WERF program effort.

1.4   Tracking Utility Consumption and Costs

An effort that needs to be a priority at a treatment facility is tracking the facility's energy consumption and cost. This can be accomplished by assigning the task to the facility's energy advocate. The result of this effort can be used as a tool to regularly monitor, assess, and identify if the energy consumption for each treatment unit process per unit of loading is consistent or varying. This should be done on a monthly basis. If it is varying, it is then necessary to determine if the per unit cost of electricity is changing, if the amount consumed per unit is changing, or if the time the energy is being consumed is not being monitored and controlled, in which case, on-peak rates are being charged rather than off-peak rates.

In addition, monitoring on a monthly basis provides data to compare to monthly consumption values and charges being received from the electric utility. In other words, your monthly checking provides concurrence with the charges you are receiving. If the comparison renders an inconsistency, you can approach the electric utility to verify consumption and charge values.

1.5   Communicate Value of Energy Awareness

Making decision makers aware and knowledgeable of the value that is embedded in energy efficiency is a must for any project or program to succeed. In order to do this, communications to decision makers must include an assessment of the facility. The assessment describes the energy efficiency opportunity as well as the cost of the modifications to implement that opportunity and a forecast of the probable payback time. The required values presented should include

- A simple description of the identified opportunity,
- An assessment to present the financial impact of the proposed modification,
- A description of what is required to acquire the savings,
- A further description of how the modifications will result in savings over time, and
- The basic payback time for the modifications.

It also is desirable to present examples where the proposed modifications are installed at a similarly sized facility or where a configuration similar to that proposed has been used. The examples should include the savings or energy consumption being realized to illustrate savings.

Communications at the management level is also required. Management must communicate with the operations and administrative staff. Management should convey to staff the value of the savings in energy efficiency and what it can mean to the overall financial well-being of a facility. The everyday, online individual has the best awareness of how the existing system is functioning and where the best locations or unit processes are available to respond to energy efficiency modifications. Staff will need to be encouraged and recognized for the value they can contribute to the facility.

2.0 ENERGY-EFFICIENT DESIGN

2.1 Energy Consumption Minimization

A facility needs to develop a target response to energy efficiency. The value of the target will vary with the knowledge and awareness the staff and, particularly, management has of energy. To develop a target, it is first necessary to identify where you are today (i.e., develop your energy baseline) and then assess and determine where you want to be tomorrow (i.e., develop your benchmark). This should begin with all facility staff becoming aware of the consumption of energy by the facility.

This process will begin with conceptually dividing the treatment facility into systems and then into process units, with the final breakdown to each piece of equipment. Initially, many approaches focused on equipment and identifying the most efficient motor to address and solve problems. However, this approach is not correct for "facility" efficiency. Facility efficiency is an integrated analysis to assess each system, then each process, and then each machine. Facility efficiency identifies the array of interconnections that exist as well as the level of interconnection of efficiency among all the parts to get to the value that the whole facility can achieve. This process will lead to the best identification of savings. This will also provide operators with a baseline on the energy consumption per treatment system at the facility. This data then needs to be processed to determine what the facility benchmark values are and what the target values should be.

Both terms presented here, *baseline* and *benchmark,* are presented elsewhere in this document with information on their definitions, how to determine them, and their correct use when comparing energy consumption values and establishing goals. *Baseline* is generally the energy you are presently consuming per a predetermined unit, whereas *benchmark* represents the best value your process should be able to obtain with your facility and unit treatment processes.

2.1.1   Understanding Your Energy Bill

Your monthly electric bill is a valuable tool that, on a monthly basis, presents you with data that is important to understand and monitor in order to assist in operating your facility in an energy-efficient manner. The initial mandatory action is distributing the electric bill to the personnel responsible for operation of the facility. What is further necessary is to provide instruction to all facility personnel on how to correctly interpret and use the many pieces of data that are regularly presented on the bill in order to assess the facility's consumption of energy.

Facility staff need to become familiar with and knowledgeable of the impact of on-peak and off-peak demand (kilowatts), on-peak and off-peak consumption (kilowatt-hours), and the time periods for on-peak and off-peak. Further, staff members need to be able to review the regularly monitored and recorded 15-minute demands to further understand the impact of equipment operation on the amount of energy consumed, when it is consumed, and the cost. With this gained awareness and knowledge, the staff will have additional tools to become aware of how a change in operation can impact energy consumption and cost.

2.2   Design Approach to Acknowledge the Value of Energy Efficiency

Most if not all states that have design codes do not have any reference to energy efficiency. Designers follow these documents because design documents must meet code requirements to be approved by the regulatory agency having jurisdiction where the facility is to be constructed. However, the designer should become aware of energy efficiency and incorporate its value into new designs.

Associated with design procedures is the need to develop a facility plan or, minimally, a document that presents comparisons of options. The plan or document typically concludes with a present-worth cost-effectiveness analysis to determine the option to implement. However, it does not incorporate the review and assessment of energy efficiency into the decision-making process. A life cycle cost analysis, incorporating an assessment of the energy efficiency of the system designed, is needed to choose the option to implement.

Equipment should be selected to meet present day demands effectively and in an energy-efficient manner as well as at projected design conditions. Therefore, design approach modifications need to be considered so energy efficiency receives its acknowledged value. Present standard design approaches need to be adjusted to include energy efficiency so all designs are energy efficient.

Another approach, at least for larger facilities, is to include energy efficiency as a component of the value engineering services that would value engineer design documents. A third and self-disciplining approach is for each designer to review the proposed improvements and self-test if the proposed system, as designed, can effectively meet the performance requirements while being energy efficient throughout the life of the project. The designer should ask: Are the structures that are proposed sufficiently flexible in size that they are efficient to operate under initial startup conditions as well as at design conditions?

Equipment selection also must be addressed. The treatment capability of the equipment selected is determined through the application of in-place codes; however, the designer must assess what combination of equipment selection will be of a size to be energy efficient when the improvement initiates operation and meets design conditions. This conscientious review of equipment selection is extremely important because many facilities never grow to their forecasted design conditions, resulting in the improvement continuously wasting energy.

2.3 Energy Efficiency Education

Implemented energy efficiency opportunities at a facility can result in a 20 to 30% reduction in energy consumption. Therefore, energy efficiency education is a much-needed action for facility staff, irrespective of the size of the facility.

To guide energy efficiency education, Wisconsin Focus on Energy has developed the *Energy Efficiency in Water Wastewater Guidebook*. The U.S. Environmental

Protection Agency (U.S. EPA) recently published an energy efficiency guidebook, *Ensuring a Sustainable Future: An Energy Management Guidebook for Wastewater and Water Utilities,* that is also available (http://www.epa.gov/waterinfrastructure/pdfs/guidebook_si_energymanagement.pdf). Some universities have professional development courses available on energy efficiency as well. Available educational programs are expanding in this area in response to the value being realized by facilities that have implemented modifications.

2.3.1   Life Cycle Cost Analysis

Life cycle costing analysis is a method of analyzing the cost of a system, piece of equipment, or product over its entire life span. Life cycle costing enables you to define the elements included in the life span of a system or product and assign equations to each component. It is necessary to use a formula that includes the value or cost of energy involved or consumed in the operation of the equipment or process being assessed.

The objective of performing a life cycle costing analysis should be to choose the most cost-effective approach for using available resources during the entire life span of the equipment, process, or product. The process should be used to assess the particular characteristics of the system over time, including energy consumption to perform the work specified.

A reference for life cycle cost analysis is presented through U.S. Department of Energy's (DOE's) Executive Order 13123, which is available at http://www1.eere.energy.gov/femp/pdfs/lcc_guide_rev2.pdf. Another life cycle cost analysis approach is available through the Hydraulic Institute (Parsippany, New Jersey) in their 2001 publication, *Pump Life Cycle Costs: A Guide to LCC Analysis for Pumping Systems.*

2.4   Gather Data

The first step in preparing an electrical energy management plan is to conduct a W/WWU survey in which all W/WWU electrical equipment is inventoried. Because a survey can be provided at different levels of detail, the reader should consider the level of survey necessary for the facility: Level 1—Walk Through, Level 2—Audit, and Level 3—Operational Simulation. Again, the designer should use the survey level that best meets the needs of the facility. Survey data to be collected include

- The name of the equipment;
- Nameplate information, such as motor power and revolutions per minute;

- A description of whether the motor has a constant load, is operated full time or part time, is of constant speed or variable speed, or has a variable load; and

- Other data relating to heating, ventilation, air conditioning equipment, outdoor lighting, disinfection system, laboratory equipment, and other factors.

A good portion of this step may already be available in many W/WWUs through their maintenance management system.

Gathering data consists of taking inventory of the W/WU's electrical equipment and listing nameplate data on all electric motors of at least 745 W (1 hp) and other equipment of at least 1 kW. During the W/WWU survey, it is typically best to start at each motor control center (MCC) and itemize each piece of equipment in order, as listed on the MCC. Also, all electric meters on MCCs and local control panels (such as kilowatt meters, ampere meters, and equipment hour meters) should be itemized.

The appropriate design data for a driven element, such as the capacity of a pump or blower, as well as motor efficiency can be obtained from equipment submittals, which should be on file at the W/WWU. Determine whether a pump or blower is associated with a flow meter, pressure gauges, an ammeter, or other equipment.

Information should be gathered from operating and billing records. Operating records along with runtime meters can be used to determine frequency of use for W/WWU equipment and create an historical comparison of energy use. Usage records from the local electric or gas utility should also be collected. The local utility will provide a summary of electric use and demand for each billing period over the last 12 months at no charge upon official request.

A qualified electrician should check the power draw of each major piece of equipment. Appropriate readings would include ammeter, voltmeter, wattmeter, and power factor readings (as equipment is available) for large motors under full load as well as partial load for variable output equipment. Flow meter and pressure gauge readings, as available, should also be made, along with the various electrical readings. An instrument mechanic should verify the accuracy of all panel meters.

Data gathering is an endless task. Operator log sheets should be checked to determine whether operators are recording appropriate energy management information, such as electric meter totalizer readings, runtime readings, and ammeter readings, at least once a week. If this information is not on log forms, forms should be developed for listing the various meter readings not being recorded. In most cases, it may be appropriate to create a new log sheet for entering energy management data. It is best to organize entries in the order in which the new data will be collected; for example, in the order of entry on MCCs (if hour meter readings will be taken from there). Data can be collected daily, weekly, or monthly. Weekly data collection may be preferred because it smoothes out daily aberrations and potential problems can be detected early. In addition to electrical readings and hour meter readings, data to be collected include influent flows, return activated sludge flows, blower supply, and analytical data, including mixed liquor volatile suspended solids and biochemical oxygen demand (BOD) data for appropriate wastewater streams. While BOD data may not be as timely as other information, it is still worthwhile to compare past aeration system electric use with actual wastewater treatment plant (WWTP) loadings.

The challenge with this is to choose the data to record and then record it. Therefore, any steps that can be taken to automate the data gathering process increase both the effectiveness and efficiency of data gathering. Instrumentation and software to monitor, record, and report on energy consumption should be incorporated with all improvements during the design phase.

2.4.1   Treatment Process Sub-Metering

The challenge associated with the majority of energy surveys and audits is identifying the amount of energy each piece of equipment is consuming. This further complicates forecasting the amount of energy used by each treatment process. Therefore, it is necessary to arrange with the owner to have a qualified and trained person or firm install an individual monitoring meter device on each piece of equipment to be assessed. It is recommended that monitoring equipment be a recording watt meter that will monitor a piece of equipment for a minimum of a week, and preferably a month, to obtain a representative sample of the power draw of the piece of equipment. If this is not available, then the person performing the survey must use the best estimating tools to forecast the amount of energy used by each process because the normal data provided by an owner is the monthly energy bill, which represents the total amount of energy used by the total facility, not what each process consumes. More accuracy is obtained through individual monitoring or sub-metering than by forecasting the distribution of the monthly consumption value received from the owner.

2.4.2   Develop a Baseline Consumption for Your Facility

It is essential to identify how much energy a facility is consuming as it exists and operates. This establishes the *baseline*, which is the amount of energy consumed by the facility to treat its waste as it is being operated. This is the value to work from to become energy efficient. The initial action is to establish this value as the baseline, and then survey the facility for opportunities to save energy and forecast the value of savings (consumption and cost).

2.4.3   Develop a Benchmark Consumption for Your Facility

This is a value that must be developed for an individual facility. The value should reflect the minimum amount of energy the facility's unit treatment processes need to treat the influent to meet its effluent limits. Once this value is established, it should become the energy consumption goal for the facility. Benchmarking data is available at a number of locations, including Wisconsin Focus on Energy's *Water and Wastewater Energy Best Practice Guidebook* (http://www.werf.org/AM/Template.cfm?Section=Home&TEMPLATE=/CM/ContentDisplay.cfm&CONTENTID=8541) (Table 3, Best Practice Benchmarks and Top Performance Quartiles for Wisconsin Wastewater Facilities) and Pacific Gas and Electric Company's (San Francisco, California) *Energy Benchmarking Secondary Wastewater Treatment and Ultraviolet Disinfection Processes at Various Municipal Wastewater Treatment Facilities* (http://www.cee1.org/ind/mot-sys/ww/pge2.pdf).

2.4.4   Data Trending

Utilizing data that has been gathered can take many forms. One comparative method is to develop trend graphs to view the consistency of the information gathered. Trends provide the viewer with a pictorial representation of loadings (BOD, total suspended solids, million gallons per day, kilowatts, kilowatt-hours) and other important parameters to monitor. Additional trends can be developed and monitored (kilowatt-hours/million gallons, kilowatt-hours/pound of BOD) to observe how the facility energy consumption varies throughout the day, daily, or weekly. The trends will allow a quick view and comparison to identify if changes are occurring and provide data to compare to other facilities' operations.

2.5   Analyze Data

Once the raw data are collected, they must be properly analyzed. One of the first steps to electric energy management is creating an electrical budget using historical information. With the data collected, estimates of usage for the past measurement period are chronicled and compared with the electrical use recorded on the W/WWU wattmeter. Using the budget analogy, the wattmeter can be thought of as measuring income ("incoming") and the individual equipment usage as expenditures ("outgoing"). Once this budget has been adequately tested to verify where all the electrical energy is being used, one can determine where and how electrical energy use can be controlled or reduced. The budget approach should be continuous, comparing weekly usage with forecast usage.

There are various approaches to estimating actual electrical use in the absence of wattmeters on all pieces of equipment. For constant-speed equipment, it is often easiest to use runtime meters and power draw for each piece of equipment. For pumps and variable-load equipment, runtime meters are seldom of any use and, as such, other estimating methods must be used.

2.5.1   Pump Systems

Pump systems (refer to Chapter 4) consume energy in relation to the amount of fluid pumped through the system in which they are installed. The primary goal is to assess the "pump" as a system, rather than just a single piece of equipment. The need is to assemble all the data available on the system (pump, motor, drive, piping configuration, and performance requirements). Then, using the information collected, the designer should use the information provided in this manual under Chapter 3 Electric Motors and Transformers, Chapter 4 Pumps, Chapter 5 Variable Controls, Chapter 8 Aeration Systems, Chapter 9 Blowers and Chapter 10 Solids Processes to assess the "system" in order to identify which component of the system is the least energy efficient and can be modified to make the entire system more energy efficient. However, the performance of the system must meet the variability of the process requirements.

2.5.2   Aeration Process

Aeration equipment (refer to Chapter 8) typically represents the greatest energy-consuming items in a WWTP (typically 50 to 70%). The energy consumption of aeration equipment is difficult to estimate. The processes' energy consumption is a variable that can be related to the BOD loading to the aeration system and to the food-to-microorganism ratio under which the system is being operated (refer to Chapter 8 for further discussion). It also needs to be checked or compared to other physical constraints that are associated with an aeration system design. The energy efficiency parameter must also be checked with the minimal airflow rate required for mixing and the airflow rate required for the number of diffusers present in the aeration system. The airflow also must be checked against code requirements to provide a certain airflow rate. After a facility has reviewed and assessed all of these measures, a decision can be made relative to the range of airflow rate that is needed because the ability of the system should be the minimal air required for low-flow conditions at the day of startup and the ability to meet conditions at design day (peak day, 20 years in the future). The system should have the ability and flexibility to meet existing conditions and future conditions in an effective (i.e., meeting effluent requirements) and efficient (i.e., using the minimal value of kilowatt-hours/unit of loading) manner. Per-unit loading parameters that could be used are kilowatt-hours/million gallons and /or kilowatt-hours/pound of BOD. Values for these parameters are available in

Wisconsin Focus on Energy's *Water and Wastewater Energy Best Practice Guidebook* (Table 3, Best Practice Benchmarks and Top Performance Quartiles for Wisconsin Wastewater Facilities), Pacific Gas and Electric Company's *Energy Benchmarking Secondary Wastewater Treatment and Ultraviolet Disinfection Processes at Various Municipal Wastewater Treatment Facilities*, and publications available through the Electric Power Research Institute (Palo Alto, California).

2.5.3   Solids Handling Process

Equipment used in the thickening, dewatering, and stabilization of sludge (refer to Chapter 10) is often operated intermittently in relation to the amount of sludge to be processed. Often, solids system equipment consists of many small motors for drives, conveyors, pumps, and so on. It is typically best to record runtime for the system and use a summary of the system's connected power.

2.6   Developing an Energy Management Plan

2.6.1   Implementation Plan

Energy management plans can take on many different forms. A comprehensive and effective energy management plan should include the general and customized characteristics listed under Section 1.1 (entitled "Energy Management Plan") at the beginning of this chapter. However, the most important feature of any plan is the implementation stage. Audits, surveys, and studies of energy use should pinpoint areas of both high energy use and inefficient energy use. Plans may include specifics of capital projects to recommend and fund, but they should also provide for continuous monitoring of use and for a means to see when inefficiencies arise. Continuous monitoring and assessment of new technologies that may result in improved energy efficiency also need to be incorporated into keeping the plan current.

2.6.2   Modify Operations

2.6.2.1   Peak Electric Demand Reduction

As discussed in Chapter 2, an electric bill for a typical WWTP consists of four major components: customer charge, demand charge, energy charge, and fuel surcharge. This chapter focuses on the demand charge portion of the bill as it relates to peak electric demand reduction. The other three aforementioned components are explained in Chapter 2.

Demand is defined as the rate of using electrical energy. For example, 10 100-W lamps burning at the same time require 1000 W (1 kW) of electricity-generating capacity or demand. Burning for 1 hour, the lamps would consume 3.5 MJ (mega-joules) (1 kWh) of electric energy.

Demand charge is the charge billed to a customer for the utility company to main-tain the generation, transmission, substation, and distribution capacities needed to meet the maximum demand. *Maximum demand* is the simultaneous demand imposed on these facilities owned by the utility company from the different users during a given time interval. The demand charge is based on a rate set by the utility company in its rate structure. It is multiplied by the demand reading as measured and as defined in the rate structure.

The *demand reading* is the highest electrical demand registered in a set time interval (typically 15, 30, or 60 minutes) during the billing period. Demand reading is measured either in kilowatts or kilovolt-amperes (kVA). In billing on a kilovolt-ampere basis, all power used, including nonworking power, is measured and billed, while billing on a kilowatt basis only includes the working power. Nonworking power is the power required to produce magnetic fields needed for operation of any inductive load, such as motors or transformers. A surcharge may be levied for a low power factor when using the kilowatt basis. The difference between the kilowatt basis and the kilovolt-amperes basis is related to the power factor, as defined in Chapter 2. A low power factor indicates a large difference between kilovolt-amperes, kilowatts, and wasted electrical power. The objective should be to maintain the highest power factor possible for high electrical efficiency. Smaller demand capacity customers are typically charged based on kilowatts. Larger demand capacity cus-tomers are typically charged based on kilovolt-amperes. This discussion will use *kilo-watts* because that is the most common term for wastewater treatment systems. Actual demand results in the aforementioned demand reading.

The demand charge is designed such that the customer pays a fair share of the utility's fixed investment in production, transmission, and distribution equipment required to meet the customer's maximum requirements. Typical single-family resi-dential meters do not record kilowatt demand; they record only kilowatt-hour (megajoule) consumption. Electric heat residential customers, however, may be required by the utility to measure demand. Again, the convention is that the higher the impact on the utility demand base, the more charges there are for demand-based rate application.

A utility company bases its demand charge on the highest amount of electricity con-sumed by the customer during the demand measurement interval; the more electricity used at any given time, the larger the possibility of an increase in a utility company's investment in generation, transmission, and distribution systems. One way of obtaining a reasonable return on these systems would be to average all generation, distribution, and transmission equipment costs among all customers. That approach would not be fair, however, because those who use the utility's equipment at a steady rate would be subsidizing those who do not. Although the peak demand periods are typically of a short duration, just like W/WWU peak loadings, the reserve capacity must be available.

Demand is not *instantaneous* demand, as many people assume. Starting up a large motor does not result in peak demand for calculating demand charge. Demand is the average power required during an established demand interval. The most commonly used demand intervals are 15 minutes and 30 minutes, but 60 minutes and other time intervals are used by some companies. The typical demand register records average energy consumption for each 15- or 30-minute interval in a day. When the first interval ends, the equipment resets and starts on the second one. The utility reviews the demand records at the end of each billing period. In most cases, the maximum demand recorded is used to compute the demand charge. There also is a ratchet clause in some rate structures that states that charges are to be based on the highest demand achieved over a period of six or 12 months, even if the present month peak demand is lower than the previous peak demand. Utility companies having problems meeting the system-demand requirements will generally devise the rate structures with a ratchet clause. These ratchet clauses are what make peak demand control cost-effective. In some instances, the ratchet clause applies only to a certain percentage of the previous peak demand. In some instances, peak demand is measured according to on-peak and off-peak hours.

Methods of demand charge and rate calculation are unique to each power company and must be understood before a program aimed at peak demand reduction is initiated. Generally, power companies are willing to explain the billing methods and will probably offer suggestions for reducing the demand charges. Utilities can often supply a demand record by intervals upon request. An analysis of this record will identify the periods of peak demand. An investigation of operations during peak periods typically reveals activities that can be deferred until nonpeak periods. For example, testing of backup pumps and blowers during a peak-load interval is expensive. Money could be saved by testing and using electric equipment with high-energy demand during off-peak times. Another method is to use an alternate power source or storage facility for storm water during the peak demand period.

Figure 11.1 presents a graph of months versus energy consumption and peak demand for the La Crosse, Wisconsin, wastewater treatment facility. The graph has a peak demand occurrence range of 810 to 884 kW ([***APPROX***]~10%). Consumption has a range of 440 000 to 590 000 kWh (34%). Note that the variation in consumption and demand values vary, but do not occur during the same months. Consumption and peak demands do not occur over the same range nor do they occur at the same time. Plots of previous years of electric consumption that are similar to this should be reviewed to identify how the WWTP has been operating and to identify minimum demand and consumption targets.

In predominantly peak-causing billing, in which frequent simultaneous starting and stopping of equipment occurs in lieu of continuous running of equipment, the demand charge portion of the bill could easily exceed 50% of the total charges. Demand charges generally make up 25% of the total electric bill for a typical W/WWU.

While the aforementioned power company rates were used in developing the models for determining demand/energy consumption and cost, it is important to

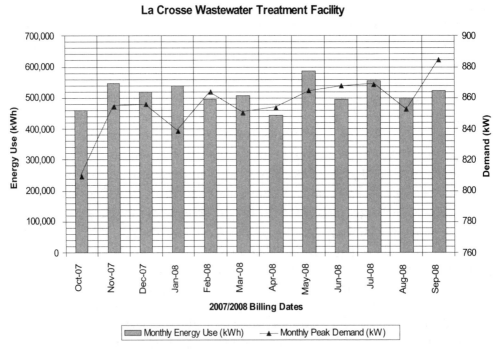

FIGURE 11.1   Months versus peak demand at a wastewater treatment facility (courtesy of SAIC).

note that the rate structures are constantly under review and are changed often. For example, the power company could examine the effects of changing the 60% ratchet to 80 or 100% in a given rate schedule. For this reason, the effects of rate changes at different locations should be examined on a regular basis.

In conclusion, the three major methods of peak electric demand reduction being presented here should be implemented after thorough investigation of the existing facility because the initial design capacity was oversized to allow for peak demands.

2.6.2.2 *Flow Equalization*

2.6.2.2.1 Wastewater.   Equalization basins provide one method of equalizing the change in demand loading on a plant's electrical system. Equalization basins are sized according to the diurnal variations in loading for a specific facility. Their volume is generally less than 35% of the WWTP capacity. Equalization basins can be located on-site or upstream of the WWTP and arranged as in-line or off-line tanks. Off-line basins or tanks, which are most commonly used, are constructed of a variety of materials. Lined earthen basins are one common form. At partially loaded WWTPs with design capacities intended to provide service to a growing community, some part of the original tanks can probably be used for flow equalization, with only minor modifications. The opportunity also exists at upgraded or expanded facilities that have abandoned structures to convert unused tanks into equalization basins, with minimal capital expenditures.

Aeration and/or mixing may be required to avoid septic conditions or settling. Generally, this entails a low power demand, but must be included when considering flow equalization.

When provided to dampen the effects of storm flows, equalization typically affects only raw wastewater pumping. The effect is typically not large and depends on the amount of runoff and presence of infiltration into the sewers or manholes. Analysis may need to be made to rehabilitate sewers and manholes versus using one of the peak-shaving methods described herein. Equalization can also be valuable in minimizing aeration energy requirements by reducing the peak daily load and air requirement. In certain applications, this benefit may outweigh pumping savings because lower maximum daily airflow rate increases oxygen transfer efficiency and reduces demand charges responsive to lower peak BOD loading.

It is more difficult to flow equalize wastewater. Some large cities have observed that the unused volume of major interceptors can be used to equalize peak daily flows without risking backups into customers' basements. An alternate to this is some cities adjust wet well elevations to provide additional storage in the wet well and use a portion of the collection system as storage. However, this approach must be carefully assessed to ensure no basements are subject to backups. Some WWTPs have installed equalization basins to provide off-line storage for incoming flows. Systems like this must be carefully designed to avoid offsetting time-of-use charges with increased pumping costs. The normal situation will be that flows must be treated as they arrive at the plant. An example of influent flow equalization at a 38-ML/d (10-mgd) WWTP was described by Porter (2007).

2.6.2.2.2 Water.   Flow equalization can be practiced in water supply and distribution through the use of ground storage tanks and timed operation of pumping operations. This is of interest for those systems that have time-of-use charges as part of the electric rate system. With time-of-use charges, rates are higher during certain parts of the day. Timing pumping operations to fill storage before the high rate period begins is one way of reducing energy costs. Timing pumping to avoid periods (if possible) of peak demand will reduce the risk of increased demand charges. However, water must be supplied if needed.

2.6.2.3 Priority Load Shedding
Priority load shedding reduces peak demand by turning off or keeping off loads that are non-critical to the process or operation of the system during peak demand periods. Priority load shedding can be achieved by manual means or by using automatic control systems.

For example, wasting of sludge, if it is not detrimental to the process, can be accomplished during the nonpeak or nighttime operation, when peak demand charges or ratchet clause effects are smaller. In addition, aerators or aeration blowers can be used in cycle fashion in lieu of turning them all on and running them together.

Scheduling of various loads in order of priority with shedding of non-critical loads is called *priority load shedding* or *load management*. Load management is the scheduling or control of electrically powered equipment to minimize peak electric demands. The two techniques addressed in this section for meeting this objective involve optimizing W/WWU operations, scheduling, and establishing a set of actions to take if demand exceeds a predetermined level.

Once W/WWU operation schedules have been optimized, a plan of action should be developed to further reduce electrical demands that are contingent on the actual demands experienced at the W/WWU. This particular aspect of load management is often automated to avoid operator errors, provides greater vigilance, and frees the operator for other tasks. However, automation is not essential. A wide range of automated load management (or energy management) systems are available. Some of the major system types and their functions are addressed here. A growing trend is toward the use of systems that not only automate load management functions, but control and schedule major W/WWU operations as well. Automatic systems can be stand-alone systems, available in the market from numerous manufacturers, or they can be incorporated in an overall computer control system for W/WWU process control. Features will vary from manufacturer to manufacturer and model to model; however, the factors discussed in this chapter should be considered when purchasing or specifying the equipment.

Feedback signal for these systems can be measured by two different methods. Separate current transformers to measure current and potential transformers to measure the voltage can be installed at a main service location to calculate kilovolt-amperes using the following formulas:

For kilovolt-amperes-based rate structure,

$$kVA = \frac{\sqrt{3} \times \text{Voltage} \times \text{Current}}{1000} \tag{11.1}$$

For the kilowatt-based rate structure, the power factor must be measured using a power factor transducer. The following formula should be used for kilowatts:

$$kW = \frac{\sqrt{3} \times \text{Voltage} \times \text{Current} \times \text{Power factor}}{1000} \tag{11.2}$$

The second approach, which is the most preferred, less expensive, and most accurate, is to ask the utility company servicing the location for the pulse input from their metering system. This pulse input, with appropriate multipliers, can be connected directly to the load management system. This input will represent actual kilovolt-amperes or kilowatts being metered by the power company.

In addition to understanding the basic functions of load management systems, it is important to know the basic steps in developing a load management plan. The following steps are recommended to help develop such a plan:

1. Determine the electrical load imposed by various pieces of equipment and rank them by importance and ability to be cycled.
2. Establish electrical demand goals and demand levels to trigger corrective action.
3. Provide some means by which to compare actual online electrical demands with the goals and targets established under item No. 2 above.

4. Specify a sequence of actions to be taken if demand trigger thresholds are reached.
5. Determine the amount of time the load can be kept off.

Only when these steps have been accomplished can the decision and design of automated equipment be determined. It should be noted that load management systems are only applicable when the facility is demand-metered and has numerous discretionary loads (loads that can be turned off).

There are a number of calculations involved to determine the cost-effective application of automated load management systems. It is advantageous to use spreadsheet-type software packages to develop a customized spreadsheet. Different rates can be applied to the spreadsheet to arrive at data that will enable a decision to be made for a comprehensive load management system.

Priority load shedding or load management involves setting up a contingency table in which various pieces of equipment are ranked in order of expendability. The most expendable pieces of equipment can then be targeted for "turn off" as demand levels reach the appropriate trigger point.

Data for maximum shed time, minimum shed time, minimum restore time, and the priority number are all derived from design and operation experience of the particular W/WWU. These values are not calculated; rather, they are determined from the answers to the following questions:

1. Which process equipment is least important to the process?
2. Which process equipment is most important to the process?
3. What is the maximum and minimum amount of time process equipment can be operated without adversely affecting W/WWU operations?
4. The W/WWU cannot run without what equipment? This equipment should not be connected to the load management system.

Horsepower or kilowatt data, which are either design data or data from the nameplate of the equipment, are important in determining the total kilowatt amount being shed.

2.7 Demand-Side Management

Demand-side management is an approach for a facility to control the demand placed on the power grid by its facility. Initially, a facility needs to learn what its demand is and when it is occurring. This information is typically available through the facility's electric company account manager. A number of energy companies have demand information available on their Internet sites. A customer just needs to contact his or her power supplier to obtain a password to access the information. The information may only have a few hours' delay from being real-time values. The information available gives the 15-minute demand values used to develop the demand value that a facility is billed monthly. The information received could be a matrix of values or a graph presenting the information in a trend manner. When the information is received, it should be reviewed and assessed to identify if there is consistent consumption or variable consumption. One of the first values that should be researched and identified is the peak demand that was recorded. Using that information, the designer or operator should return to that day and time to reconstruct what may have been the cause for the high consumptive value. At times, the high value is produced by high flows because of a rain event. However, it could be the result of an additional blower being activated because of an increased organic loading or it could happen because operations is testing a system or flexing membrane diffusers. Finally, it could be due to a hauled load of material being discharged into the receiving station and added immediately to the facility, which may be the same as the peak demand period.

Therefore, it is necessary to

- Assess the power company-provided demand values and request them as a graph;
- Assess the values to identify when the peak demand occurred; and
- Reconstruct the day to identify what caused the peak demand and determine if it could be avoided through actions that could easily be put in place.

2.8 Communicate the Value of the Energy Management Plan

Communicating the need for energy efficiency improvements to elected boards and/or the public is valuable for gaining support for an effective energy conservation program. Maintaining lines of communication with local elected officials and the community will often sustain support for initial successes. Communication should be two-way: listening is an important part of communication and suggestions from staff, elected officials, and the public will also improve the effort. U.S. EPA's Energy Star Web site, titled "Elevate Energy Management to Senior Managers," also provides useful information on the topic (http://www.energystar.gov/index.cfm?c= industry. bus_industry_elevating).

Facilities with effective conservation efforts have found that motivating staff to conserve energy and providing work incentives, such as payment for ideas that improve operating energy conservation and economy of operation, are effective ways to achieve and sustain energy conservation goals. Additionally, local electric and gas utilities should be contacted to inquire what programs they have in place to promote and assist in funding energy efficiency projects.

3.0   BENEFICIAL USE OF RENEWABLES

3.1   Biogas

Biogas is a natural compliment to operations at many WWTPs that use anaerobic digestion to stabilize wastewater solids. Biogas produced by the digestion of biosolids is approximately 60% methane and can be used to power process boilers, comfort boilers, engine-powered generators, turbine-powered generators, engine-powered pumps or blowers, and fuel cells.

A few WWTPs have found it beneficial to pipe landfill gas to a site to fuel engine or turbine-powered generators. A recent trend is the co-digestion of waste food products with digesting biosolids.

An effective way to use biogas is in a combined heat and power application. All of the biogas is directed to and burned in an engine or turbine that, in turn, powers an electric generator. The electricity is used on the treatment plant site. The waste heat is recovered and used to heat the anaerobic digester, to provide heat to the buildings, or for other uses around the plant site.

The generation and use of biogas are discussed in further detail on U.S. EPA's Combined Heat and Power Partnership Web site at http://www.epa.gov/chp/.

3.2   Wind

Wind provides a viable source of renewable energy to meet much of the power needs at a water or wastewater facility. It makes sense to generate somewhat less than needed so that the entire production of electricity is used on-site. This results in all renewable power being priced at wholesale prices. (Visit the DOE's Energy Efficiency and Renewable Energy, Wind and Hydropower Technologies Program Web site at http://www1.eere.energy.gov/windandhydro/.)

Development of wind resources usually requires 1 year or more of on-site monitoring to provide sufficient evidence to obtain funding. Wind maps to assess the potential for an economic wind energy project are available through the aforementioned DOE Web site. Generally, a wind category of 3 or better is needed.

The delivery time for large wind turbines is determined by worldwide availability, and a wait time of up to 36 months for delivery has been reported. Consideration must be given to foundation requirements, long-term leases or commitment of land/site, and potential concerns by neighbors on issues such as visibility, noise, and the impact on birds. Visit the Atlantic County Utilities Authority Renewable Energy Web site for an overview of wind projects (http://www.acua.com/acua/content.aspx?id=488&ekmensel=c580fa7b_20_88_btnlink).

3.3 Solar

Solar energy is a popular renewable resource at water and wastewater facilities. These sites can use all of the power that is generated. Solar-thermal produces hot water or hot air that can be used to heat buildings or provide heat for digester needs. Solar photovoltaic produces electric power that can be consumed at on-site buildings at 120 V, or on process equipment at 460 V if an inverter is used.

Solar power is also becoming a source of power for remote sampling stations, lighting, and lagoon mixers. Visit the DOE's Energy Efficiency and Renewable Energy, Solar Energy Technologies Program Web site at http://www1.eere.energy.gov/ solar/.

3.4 Biomass

Biomass is waste wood (or another product) that has fuel value. The wood is heated and, in the process, gives off a low British thermal unit fuel that can be used in process boilers to produce low pressure steam, hot water, or hot air. In 2002, the Bioenergy Technology Subcommittee of the Water Environment Federation published a white paper titled *Biogasification and Other Conversion Technologies,* summarizing options for biomass conversion.

3.5 Hydro Turbines

The most common application of hydro turbines is for water supplies in which a pressure-reducing or pressure control valve has been inserted to control pressures in the water supply line or in distribution lines. In this case, the hydro turbine is installed upstream of the pressure-reducing valve, which should remain in the line to protect the system during times when the turbine is off-line. The economics of installation usually require that an electric supply into which the power can be fed be in close proximity to the hydro turbine. The cost of installing feeder lines to the nearest power supply can offset the benefit and make the project uneconomical.

Generally, the head and flow from a WWTP are not sufficient to generate more than a few kilowatts of electricity. This may be suitable for a remote sampling station, but will not make a measurable reduction in the amount of electricity used by the plant. The one notable exception in the United States is the Point Loma WWTP in San Diego, California, which has a flow of approximately 379 ML/day (100 mgd) and an elevation drop of about 30 m (100 ft) (visit http://www.sandiego.gov/mwwd/graphics/hydroplant.jpg for more information).

3.6 Fuel Cells

A fuel cell combines hydrogen and oxygen to produce electricity. The byproducts are water and heat. The fuel is converted to energy by an electrochemical process rather than combustion, with a result that it is clean, quiet, and efficient. The waste heat can be used to provide hot water or space heating for a home or office. Benefits include reliability, multi-fuel capability, siting flexibility, durability, scalability, and ease of maintenance (visit http://www.fuelcells.org/ for more information).

There are several types of fuel cells, with several more under development. They tend to fall in the 5-kW to 1000-plus-kW size range, including phosphoric acid, proton exchange membrane, and molten carbonate.

Fuel cells require hydrogen for operation. However, it is generally impractical to use hydrogen directly as a fuel source; instead, it must be extracted from hydrogen-rich sources such as gasoline, propane, or natural gas. Wastewater treatment plants with anaerobic sludge digesters produce methane-rich biogas, a favored source of fuel for fuel cells.

Current applications of fuel cells have been limited due to relatively high costs (e.g., $4,800 per kilowatt or more for a fuel cell). The U.S. Department of Energy formed the Solid State Energy Conversion Alliance with a goal of producing a solid-state fuel cell module that would cost no more than $400 per kilowatt (visit http://www.fossil.energy.gov/programs/powersystems/fuelcells/fuelcells_seca.html for more information).

4.0 ON-SITE ENGINE OR POWER UTILIZATION

If neither flow equalization nor peak electric demand reduction is feasible, then an on-site engine or power generator should be considered.

Most W/WWUs require an emergency source of mechanical or electrical power, as mandated by U.S. EPA. This source is required for the treatment process operation during a power interruption from the power company supplying power under normal operating conditions. The emergency source of power could be engine-driven equipment (blowers, pumps, and so on), a second electrical power feed, or an on-site generating system. The power company typically charges an exorbitant amount to provide a backup power feed. Because of this, it is generally economical to provide an on-site power-generating system. In addition, the option of using an on-site generating system for peak shaving or peak electric demand reduction could be considered. There are four different design/application methods for using on-site power-generating systems:

- Engine-driven equipment,
- Power-generating system—traditional transfer scheme,
- Power-generating system—synchronized transfer scheme, and
- Power-generating system—parallel with utility.

While considering this option, engine/generator operation and maintenance costs should be considered against the savings that could be realized from peak electric demand reduction.

4.1   Engine-Driven Pump

The engine-driven pump method is inexpensive to implement. If the engine-driven pump is required as mandated by U.S. EPA for backup purposes, then capital cost to provide the pump should not be included in the payback analysis to be performed for peak electric demand reduction. In this method, it is also assumed that the W/WWUs are manned 24 hours/day and that additional manpower to run the pump during peak electric demand periods is unnecessary.

Dual-driven systems can be used to increase overall reliability. An electric motor and a right-angle drive for the engine can alternately both drive the pump. During high-demand periods, the engine can be turned on and used to drive the pump, thus saving peak electric demand costs.

After analyzing the need for peak electric demand reduction, the pump can be turned on either manually or automatically during the peak electric demand cycle. If peak-electric-demand-causing periods are predictable, then exercising of the pump can be synchronized with the peak demand occurrence. Standby pumps typically must be exercised under load once per week.

The subsequently described methods are used more often than the engine-driven pump, mainly because of the lack of flexibility of power use and the unavailability of standby electrical power.

4.2   Power-Generating System—Traditional Transfer Scheme

In this option, a power-generating system is used. All of the aforementioned methods achieve the same goal: provide power for peak electric demand reduction. This goal is achieved at different costs and savings, depending on the method used. Along with cost, the convenience of using a system that is adaptable to a particular application should be considered. This option provides the flexibility of selecting any load in the system rather than only one that has an engine connected to it, as previously discussed.

When considering this option, the power-generating system should be designed to provide peak power for the duration that such power is necessary. It may be necessary to derate the generating unit because of the longer continuous running time required compared to the standby application.

A traditional transfer scheme will create total power interruption. It is recommended to provide dead time between power transfer from generated power to restoring line power to ensure that back electromagnetic forces from the de-energized load have been completely decayed. In-phase monitors are available as an option if total power interruption is to be avoided. This method is the safest and least complex of the three described in this section.

4.3   Power-Generating System—Synchronized Transfer Scheme

In this method, the power system grid and the generating system are synchronized and allowed to operate in parallel during the generation of power. Power is then fed into the W/WWU electrical distribution system as required by the grid demand.

The advantage of this method is that the load does not have to be turned off during the peak shaving cycle, which would cause a process interruption. A synchronized transfer scheme will keep wastewater pumps running during the transfer cycle, providing closed transition transfer, and will not take the pumping system off-line. It may be necessary to use these types of systems when the process upset is not acceptable.

4.4   Power-Generating System—Parallel with Utility

This method is similar to the synchronized transfer scheme. The major advantage with this method is that if the city or village that owns and operates the W/WWU also owns and operates the electric utility, then the electric-grid-wide peak electric demand reduction could be realized by operating the generating system. However, it is necessary to obtain acceptance from the electric utility being connected to.

On a hot summer day, when all air conditioning units are running and brownouts are occurring in the city electric utility system (because of heavy demand from the customers and with demand at the W/WWU not that high), the generating system could be turned on, parallel with the electric utility grid, and power can be exported to the grid to lower system-wide demand.

Payback could be calculated by comparing the cost of the switchgear and the equipment necessary to provide the paralleling system against the benefits realized in lower demand and related ratchet clause effect.

5.0   ON-SITE GENERATION OPTIONS

Self-generation options vary widely, but can be discussed generally in terms of the following three primary self-generation options available to W/WWUs:

- Using emergency backup generators for peak shaving,

- Installing natural gas co-generation units, and

- Using digester gas, sludge, and other byproducts of the wastewater treatment process as fuels.

5.1   Using Emergency Backup Generators for Peak Shaving

Nearly every W/WWU has backup generation capability of some type for emergency situations. Depending on the structure of electric rate options provided by the local utility, peak shaving may be a viable method of achieving significant reductions in power costs with minimal incremental investment.

The concept of peak shaving is to offset peak energy and demand purchases by self-generating during peak (highest rate) periods. This may take the form of self-generating all power required during such periods or shaving the amount of peak purchases from the local utility by supplementing peak-period purchases with self-generated power.

Because emergency generators are already installed and operational, the only additional costs typically incurred are in the following areas:

- Direct fuel and supply costs,

- Environmental permits authorizing the facility to operate at higher levels of output, and

- Staff time and training required to operate the equipment at an increased frequency.

Whereas this is an attractive approach to reducing power costs, it may not always be feasible to implement. One reason is that the local utility may not be supportive of the customer's decision to self-generate during high-revenue periods and may assess unreasonably high demand rates to attempt to compensate for the expected loss in revenues. Another problem is that standby generators using diesel fuels may not be allowed to operate on a nonemergency basis the local air pollution control district (APCD). An air permit may be required to operate beyond an "emergency" basis; therefore, the APCD should be contacted prior to implementing planned non-emergency operation. Should this be the case, the WWTP should investigate one of the other two options described in this chapter.

5.2 Distributed Generation

Distributed energy covers a wide range of small-scale power generation technologies (typically in the range of 3 to 10 000 kW) located close to where electricity is used. In the simplest form, it is electrical generation that is installed on the electrical grid where the load centers are, in contrast to the large central plants that the utility built to be close to the fuel source and away from people. Generally, it is used to discuss generation that is installed on the distribution grid at customer locations (California Energy Commission, California Distributed Energy Resource Guide, http://www.energy.ca.gov/distgen/; Peterson, 2008).

The Energy Policy Act (EPAct) of 1992 (U.S. EPA, 1992) required interstate transmission line owners to allow all electric generators access to their lines, including distributed energy resource (DER) sites. Distributed energy resources complement central power plants, and installing DER systems at or near the end user can benefit the electric utility by reducing the load on the transmission and distribution system. The application of renewable energy, such as digester gas, landfill gas, and wind, photovoltaic, and geothermal energy, are generally implemented as DER applications (California Energy Commission, California Distributed Energy Resource Guide, http://www.energy.ca.gov/distgen/).

The potential benefits to a consumer are lower power cost, improved reliability, power quality, energy independence, and the ability to obtain power in remote locations. The range of installations that a DER covers is broad, and examples of installations that may be found at water and wastewater facilities are described in the following section.

5.3   Installing Cogeneration Units

Another approach is to install one or more cogeneration units to supply the facility's energy needs with power purchased from the local utility only during emergency or down periods. The economics of this option depend on a variety of factors, including

- The size, capacity, and associated capital investment of generating units required to serve the facility's electric requirements;

- The availability of financing, rates, and terms;

- The cost and availability of fuel resources;

- Operation and maintenance costs (such as labor, equipment, maintenance, and allowance for capital repairs);

- The ability of the W/WWU to use thermal energy produced by the cogeneration equipment and the corresponding economic value for offsetting thermal energy purchases; and

- Opportunities, if any, for the W/WWU to sell excess power produced to the local utility or another power consumer, and the expected revenues produced by such excess power sales.

To assess the viability of such an option, a comprehensive cost-benefit analysis should be performed, incorporating engineering design criteria and identifying optional configurations. This kind of study should be performed by a specialist in this field who is familiar with both the technical aspects of cogeneration and the economic analysis of cogeneration projects.

5.4   Using Biogas, Sludge, and Other Byproducts of the Wastewater Treatment Process as Fuels

A variation of the cogeneration theme would entail using digester gas, sludge, and other byproducts of the wastewater treatment process as fuels. This approach can significantly enhance the economic benefits realized by offsetting or eliminating the need for purchased fuels. Additional key economic variables are:

- Capital investment,

- Financing terms,

- Cost and availability of fuel

- Energy efficiency or renewable grants from local utilities or state,

- Operation and maintenance costs,
- Value of thermal energy produced, and
- Potential revenue from sales of excess power.

Any savings realized from reduced waste disposal costs should also be factored into the analysis.

Obviously, power-generation facility alternatives can be quite complex. The optimal configuration ultimately depends on operating and design characteristics specific to the W/WWU, subject to engineering design, financing, and fuel resource constraints. Other complicating factors include constraints imposed by regulatory agencies, such as environmental permitting.

5.5   Feasibility Evaluation

An evaluation of options for saving energy costs entails an analysis of a variety of factors specific to the W/WWU's circumstance. The primary factors to consider are listed in the following subsection. It is important to note that no one factor should be considered independently of the others because they are all interrelated. The project configuration selected will ultimately be that which represents the optimal choice, given the W/WWU's objectives for this project; environmental, economic, and other constraints; and the ratio of benefits to costs.

5.5.1   Ranking of Project Objectives

Before embarking on any project, the W/WWU must identify and prioritize project objectives. These may include

- Serving the W/WWU's minimum, average, and peak load;
- Using waste products from the treatment process;
- Maximizing financial benefits (through offsets of retail power purchases and/or sales of excess power);
- Constructing a showcase project demonstrating the viability of new technology or the feasibility of reusing waste products for productive purposes; or
- Any combination of the aforementioned objectives.

5.5.2   Facility Factors

For a planned self-generation project, siting of the power-generation facility is typically based on several factors, including

- Proximity to the fuel source(s),
- Location of the load(s) being served,
- Disruptions and efficiency of W/WWU operations,
- Location of nearest commercial and/or residential communities,
- Distance from utility distribution lines (if excess power is to be sold), and/or
- Space available.

Other types of factors that may enter into siting considerations are

- *Environmental*—If the W/WWU site is near a residential area, concerns about noise, smells, and other types of pollution may affect the location of the generation plant.
- *Economic*—When more than one choice of site is available, the relative costs versus benefits of the respective options may be the determining factor. For example, all other factors being equal, if the W/WWU owner has the option of siting the generation facility near the fuel source or near the utility distribution line and these are at different locations, the most economic approach in terms of capital costs may be the determining factor.

5.5.3 *Design Factors*

The design selection process involves two basic steps: determining the appropriate system size and operating mode and identifying prime mover type.

Again, a variety of factors interrelate with the design selection decision. The most important element at this stage of the analysis is to ascertain the W/WWU's power requirements, both thermal and electric, and its demand and energy profile on a time-of-use (daily, weekly, monthly, and seasonal) basis.

The W/WWU's power requirements then provide the basis for several approaches for sizing the generation facility to

- Supply the W/WWU's minimum load,
- Supply the W/WWU's average load,
- Supply the W/WWU's peak load, and
- Optimize the use of nonconventional fuel resources (e.g., sludge, digester gas, and/or landfill gas).

In order for each facility size to be evaluated, one or more prime mover types should be selected on the basis of fuel resources available, capital cost considerations, and significant performance characteristics (e.g., conversion efficiency and emissions).

Each project configuration (consisting of facility sizes and prime mover types to be evaluated) should then be evaluated on the basis of its ability to meet technical, economic, operational, and environmental project objectives.

5.5.4 Economic Factors

An economic evaluation includes consideration of a variety of factors that lead to a decision regarding the optimal project size and configuration. Primary factors include

- Wastewater treatment plant current and projected power requirements (electric and thermal, demand, and energy);
- Cost of utility-provided power, current, and forecast;
- Value of excess power generated (if an outside market exists);
- Cost and availability of fuel resources (e.g., natural gas, diesel, fuel oil, digester gas, landfill gas, or wastewater sludge);
- Site costs (procurement, rental, and lease);
- Capital costs (including preliminary design, engineering design, permitting, equipment, construction, construction supervision, and cost of capital for the various design options considered);
- Capital cost credits (e.g., capital costs offset by constructing the project, such as the cost of air emissions control equipment for digester gas, which need not be purchased when the digester gas is used as fuel);
- Cost of air emission offsets purchased and required (if any) to operate the facility;
- Operation and maintenance costs (whether performed by staff or an outside contractor);
- Savings attainable by avoiding disposal and other costs (e.g., disposal costs avoided by using sludge as fuel);
- Capital replacement cost budget (for scheduled and nonscheduled replacements);
- Tax benefits available (such as investment credits and alternative fuel credits);
- Cost of capital (whether interest paid on project debt or facility's cost of capital inclusive of equity, on a discounted time-value-of-money basis); and/or

- Proportion of equity contribution and project risks and benefits to flow to each project participant (including facility owner, developer, and financer).

All of the aforementioned factors should be quantified and input to a spreadsheet modeling the project's expected financial performance on a long-term basis (i.e., 10 or more years, depending on factors such as the estimated project life, financing and contract terms, and quantity of fuel resources available). The project may be evaluated on the basis of

- The value of discounted net cash flows over some term;
- Whether or not it generates an acceptable rate of return for the facility owner and other project participants;
- The extent to which it is expected to achieve other objectives (e.g., environmental and technological); or
- Any other set of criteria established by the project developers.

5.5.5 Operational Factors

Operational requirements of any planned generation project must also be considered. Operational factors include

- The number and types of staff required to operate the generation facility,
- Staff scheduling and training,
- Frequency of equipment maintenance, and
- Fuel processing.

5.5.6 Environmental Factors

A variety of environmental factors must also be considered when planning a power-generation project. These vary from one geographic location to another and are subject to local, state, and/or federal regulations. Consequently, the respective environmental regulatory agencies and other local authorities should be consulted for an update on the environmental constraints applicable to the project.

Although specific requirements vary widely, the following types of environmental constraints are fairly common to all generation projects in the United States:

- *General*—U.S. EPA establishes federal guidelines for protection of the environment, which are then delegated to the respective states for implementation. These regulations have resulted in a series of regulatory approvals governing construction of new facilities such as power-generation facilities. One of the most important of these is the environmental impact report (EIR), which must be prepared by a new facility. The EIR is a comprehensive report documenting the planned facility design and the types of effects it is expected to have on the environment. Environmental impacts include

- Air quality (such as organic and inorganic emissions, visibility, and smell),
- Noise,
- Water quality (such as chemical content and temperature),
- Population density (such as effect on traffic patterns, parking, and community services), and
- Cultural, historical, scenic, and other characteristics of the proposed site.

Virtually any expected effect of any kind on the environment (flora and fauna) and the adjacent community must be documented, reported, and assessed before a project can be approved. In addition to review by one or more regulatory bodies, the EIR is also subject to public review and comment.

- *Air pollution*—Air emissions are subject to regulation by the U.S. EPA, state air resources boards, and regional APCDs. U.S. EPA establishes minimum air quality requirements for each state and air basin. The respective states then establish programs to attain and/or maintain these minimum federal requirements. States may also opt to impose more restrictive requirements to attain a higher air quality standard than that required under federal law. It then becomes the responsibility of the local APCDs to administer the state program for their specific air basins.

The following is a list of some permits required by U.S. EPA and/or APCDs for construction of a new facility, such as a power generation facility, under new source review procedures:

- *Authority to construct*—The APCD will typically require the developer to submit preliminary design plans and equipment specifications, including estimated quantity and type of pollutants emitted, given the planned fuel mix. The APCD will then evaluate the data provided by the developer in terms of whether or not the facility meets local limits on certain types of pollutants. The APCD may require the project developer to apply best available control technology (BACT) to mitigate the adverse effect of controlled pollutants on the environment.

- *Prevention of significant deterioration (PSD)*—Projects exceeding certain sizes and/or emissions constraints are subject to additional scrutiny through the PSD review process. The purpose of PSD is to ensure that new sources of pollutants do not have a significant negative effect on the air quality of the region. The PSD requires that the project developer model air emissions, taking into account a variety of site-specific factors (e.g., wind currents, ambient air temperatures, and height and points of emissions), to demonstrate that the project will not have a deleterious effect on regional air quality. In addition, all major new sources are subject to BACT.

- *Permit to operate*—Upon completion of the facility, the APCD may require emissions testing to confirm that the facility will perform within emissions limits established in the authority to construct. A permit to operate will then be issued, authorizing the owner to operate the facility, subject to certain conditions (e.g., fuel constraints, emissions constraints, and operating hours) for a specified period of time. The facility owner may also be required to perform periodic emissions testing to retain the right to operate.

A variety of environmental and other permits (e.g., building) may be required, based on local regulatory objectives and constraints. Of these, air emissions constraints tend to be the most significant when determining the feasibility of a power-generation facility of this kind.

However, in the event that hazardous wastes are being considered as fuel resources (e.g., use of landfill gas from a superfund site), the project may be subject to other permits and approvals (e.g., a health risk assessment that evaluates the risk to neighboring communities as a result of the emission of known carcinogens). The key environmental factors affecting feasibility of each project configuration must be evaluated in the context of its unique mix of resources, technology, performance, and other characteristics.

6.0   FINANCING APPROACHES

The developer can finance development, construction, and/or operations of an electric-generation plant. The following are types of financing available:

- *Construction loans*—short-term financing during construction that may be refinanced with another, usually lower interest, loan upon commencement of operations;

- *Working capital loans*—short-term financing to provide working capital during development, construction, and/or operations; and

- *Fixed-term loans*—loans of a fixed duration for some specified purpose, for example, as refinancing of a construction loan.

The specific financial structure used depends on the financing objectives and the amount of risk and equity to be assumed by the project developer. The primary types of financial structures are revenue bonds, conventional bank financing, lease financing, privatization, and joint ownership and/or development. The financial structure put in place will depend mainly on the amount of risk assumed, equity to be contributed, and benefits to be allocated to the respective project participants and financers.

6.1 Project Financing

Project financing is a method of financing capital-intensive construction projects, wherein the project and its assets, contracts, operating revenues, and cash flows are evaluated by a lender as a separate entity. The primary appeal of project financing is that the debt may be nonrecourse; that is, if the developer can demonstrate that project cash flows may be reasonably expected to support all expenditures, including operations and maintenance and debt service, the developer may not be held responsible for meeting debt service obligations during periods when project cash flows fall short of expectations. Further, because project financing is dependent on the financial viability of the project alone, the debt capacity of the developer is not affected.

Any one or combination of the aforementioned financial structures can be used for project financing. The key element is that the project economic structure be able to stand on its own.

6.2 Revenue Bonds

Most W/WWUs are owned and operated by municipalities or special districts; thus, tax-exempt bonds are the primary means of financing construction of an electric-generation plant. If the W/WWU is privately owned, the developer may be able to issue industrial development bonds or industrial revenue bonds. Industrial development bonds and industrial revenue bonds may be either taxable or tax exempt. Pollution control facilities such as WWTPs, including local power-generation facilities servicing such WWTPs, are tax exempt for industrial revenue bond purposes.

When revenue bonds are issued, the revenue stream from the generation plant is committed to repayment of the bonds. Special funds (e.g., operations and maintenance, surplus, capital replacement, and capital reserve), financial management, and reporting requirements are established by the bond covenants to ensure protection of the bondholders' investments in the plant. The specific interest rate and terms available are dependent on the debtor's financial position and bond rating as well as financial market conditions.

6.3 Conventional Bank Financing

If public debt is not a viable option, commercial bank loans may be explored. These may be in the form of secured or unsecured loans and may involve one or more lenders.

The specific terms available will vary widely with the credit standing of the developer, the financial strength of the project, and financial market conditions. In general, the following characteristics will apply:

- Rate(s) of interest, whether fixed or variable, will be specified for a stated term.

- The lender is entitled to repayment of the loan with full recourse (i.e., the borrower is ultimately responsible for debt service, regardless of whether or not the project meets projected cash flows).

- The lender does not share in project risks or benefits.

To qualify for a loan, the developer will likely be required to contribute some percentage of equity, demonstrate the overall financial viability of the planned project, and pledge securities to protect the lender in the event of default on the loan.

6.4 Lease Financing

Another option to explore is lease financing. Lease financing is a means of financing large equipment purchases and projects in which ownership of the leased property is a key issue. The full range of leasing options is complex and not appropriate for discussion here. However, the primary types of leases can be summarized in the following subsections.

6.4.1 Direct Financing Lease

The lessor provides all funds necessary to purchase the leased asset (also referred to as *nonleveraged* and *capital leases*), and the lessee assumes all of the risks and benefits of ownership. Direct financing leases have one or more of the following characteristics:

- Ownership of the leased asset transfers to the lessee at the end of the lease term;

- A bargain purchase option is included, allowing the lessee to purchase the asset at a price less than the fair market value at the end of the lease term;

- The term of lease is equal to or exceeds 75% of the estimated useful life of the leased asset; and

- Present value of lease payments, including minimum payments during any noncancellable terms, equals to or exceeds 90% of the fair market value of the leased asset, less any investment or other tax credits retained by the lessor.

6.4.2   Leveraged Lease

A leveraged lease is similar to a direct lease, but involves a minimum of three parties: lessee, lessor, and long-term lender. There are many variations on leveraged leases, but all have the following characteristics:

- The lessor assumes responsibility for repayment of the loan from the lender(s) and assumes ownership of the project. The loan is secured with a first lien on the equipment, assignment of the lease, and assignment of lease rental payments. The financing provided by the lender(s) is substantial to the transaction and nonrecourse to the lessor.

- The lessor's net investment declines during the early years of the lease term and increases during the later years of the lease term.

- Investment and other tax credits retained by the lessor are accounted for as one of the cash flow components of the lease.

6.4.3   Operating Lease

An operating lease is a means of providing financing that is not considered a purchase by the lessee (also known as *off balance sheet financing* because operating lease payments are considered rents; the associated lease obligation is not recorded on the lessor's balance sheet and does not reduce the lessor's borrowing capacity). Operating leases are defined as those that do not meet the criteria of capital or the aforementioned direct financing leases. That is,

- Ownership of the leased asset *does not* transfer to the lessee at the end of the lease term;

- The lease *does not* contain a bargain purchase option allowing the lessee to purchase the asset at a price less than the fair market value at the end of the lease term;

- The term of the lease is less than 75% of the estimated useful life of the leased asset; and

- The present value of the lease payments, including minimum payments during any noncancellable terms, is less than 90% of the fair market value of the leased asset, less any investment or other tax credits retained by the lessor.

6.4.4   Conditional Sale Lease

A conditional sale lease is a means of installment purchase financing, sometimes available from large equipment manufacturers, in which ownership is transferred to the lessee with a bargain purchase option at less than the asset's fair market value. All risks and benefits of ownership of the leased asset are assumed by the lessee. A conditional sale lease is classified as a capital lease for accounting purposes (i.e., the lease obligation must be recorded on the lessee's balance sheet as a liability).

6.4.5   Certificates of Participation

Certificates of participation are a form of public financing that have the appearance of tax-exempt revenue bonds but are structured as leases. Basically, title to the property being financed is held by certificate holders and amortized over the term of the certificates. Payment on certificates is guaranteed by revenue generated by the financed facility. At the end of the lease term, ownership reverts to the municipality or public agency that issued the certificates.

6.4.6   Tax-Exempt Leases

Tax-exempt leases are direct leases financing construction of public facilities and other qualifying facilities constructed for the benefit of the public. Interest revenues earned on tax-exempt leases are not taxable. When tax-exempt debt is used, any available tax credits (such as investment, energy, and nonconventional fuels) may have to be foregone. Therefore, benefits gained from tax-exempt debt must be evaluated against the cost of tax benefits foregone, if any.

6.4.7   Sale-Leaseback

In a sale-leaseback, the owner of the asset sells the property to a third party (lessor) who immediately leases the property back to the owner (lessee). The leaseback may be either a capital lease or an operating lease. If it is a capital lease, project financing will result.

6.4.8   Lease Financing Considerations

There are many variations on project lease structures. One example of a project lease financing structure requires that the site owner obtain temporary construction financing. Upon completion of construction, the lessor purchases the project for some agreed-on price, with the project revenues committed to repaying the lease payments. Generally, the lease terms will also contain provisions intended to protect the lessor's investment by providing for, for example, the establishment of special funds for capital repairs and replacements during the term of the lease. Any excess project benefits may be shared on some agreed upon formula with the site owner or taken in full by either party.

The merits of lease financing vary according to the relationships and responsibilities of the respective parties. In general, advantages include

- Transfer of tax benefits from lessee to lessor (capital leases),
- Nonrecourse debt for lessor (leveraged leases),
- Off-balance sheet accounting (operating leases),
- Lower interest (versus conventional bank financing),
- Flexible structure for joint ventures,
- Fixed-fate payment stream, and
- 100 percent financing available.

Disadvantages include

- Loss of residual (operating leases, when asset has substantial useful life on expiration of lease term),
- Higher cost (when tax benefits could have been used by lessee), and
- Fixed senior obligation against project.

6.4.9 Energy Services Contracting

Energy services contracting is an alternative project delivery process in which savings brought about by an energy conservation project, renewable energy project, or other facility improvement are used to pay for the cost of the capital improvement.

Energy services contracting enables governmental agencies to make infrastructure and facility improvements by reducing operating expenses and making a positive impact on capital budgets. By implementing an energy conservation method (ECM), operating costs are reduced, thereby reducing waste. This allows projects to be funded without requiring tax increases, bond issues, or upfront monies from capital budgets. Additional information is available from the Energy Services Coalition (ESC) at http://www.energyservicescoalition.org/.

A municipality or other governmental agency would implement an energy services contract by working with a qualified energy services company (ESCO). Energy services companies typically act as project developers for a wide range of tasks and assume the technical and performance risk associated with the project. According to the National Association of Energy Service Companies (Washington, D.C.) (National Association of Energy Service Companies Home Page, http://www.naesco.org), ESCOs typically offer the following services:

- Develop, design, and arrange financing for energy-efficient projects;

- Install and maintain the energy-efficient equipment involved;

- Measure, monitor, and verify the project's energy savings; and

- Assume the risk that the project will save the amount of energy guaranteed.

These services are bundled into the project's cost and are repaid through the dollar savings generated. Energy services companies typically track energy savings according to sanctioned engineering protocols. The National Association of Energy Service Companies can provide additional information on energy performance contracting as well as guidance in undertaking an energy performance contract. It is also a source of information for locating qualified ESCOs. (For more information, visit their Web site at http://www.naesco.org.)

Energy services contract laws for the local state should be checked for specific requirements. However, according to ESC, working with an energy services contract should entail the following:

1. Determine if an energy services contract has merit for the specific opportunity.
2. Select an ESCO based on qualifications.
3. Enter into an agreement for the ESCO to identify energy saving opportunities.
4. Negotiate a long-term contract to implement ECMs.
5. Verify savings and enjoy the benefits.

6.4.10   Utility Services Contracting

A governmental agency may also implement energy conservation or renewable energy projects through a partnership with the utility that provides electric power through an agreement called a *utility energy service contract (UESC)*.

Under a UESC, the utility commonly arranges financing to cover the capital costs of the project and is then repaid over the term from the savings generated by the

ECM. This arrangement allows a governmental agency to implement an ECM with no initial capital investment.

According to DOE's Federal Energy Management Program (http://www1.eere. energy.gov/femp/financing/uescs.html), more than 45 electric and gas utilities have provided project financing for energy and water efficiency upgrades at federal facilities.

6.4.11 Grants/Rebates

A variety of grants and rebates are available from state energy offices, energy utilities, and federal agencies. The Database of State Incentives for Renewables and Efficiency (http://www.dsireusa.org/) provides a resource for checking on local opportunities. The database provides a comprehensive source of information on state, local, utility, and federal incentives that promote renewable energy and energy efficiency. The Web site is maintained by the North Carolina Solar Center, which is part of the College of Engineering at North Carolina State University (Raleigh, North Carolina).

6.5 Privatization

Privatization entails development, ownership, and operation and maintenance of the electric-generation facility by an independent developer, and the sale of output from the facility to the W/WWU. For example, the W/WWU may be unwilling to undertake development of an electric-generation plant because of high risk, capital costs, and/or other concerns. An outside developer may be obtained to evaluate and construct a suitable generation facility that meets the W/WWU's power requirements, is compatible with W/WWU operations, and produces sufficient economic returns to the developer.

6.6 Joint Ownership and/or Development

Between sole development and ownership by the W/WWU owner and privatization, there are a wide variety of project structures that can be used, depending on the extent of financial risk and rewards assumed by the respective project participants. Key negotiating points include

- Economic-value-attributed digester gas provided by the WWTP to the project as fuel;
- Economic-value-attributed use of the project site (i.e., rental or lease charge); and
- Ownership of project equipment.

Negotiable contract points to consider include

- Commitment to specific levels of service, for example,
  - Thermal quantity and quality,
  - Electric quantity and quality,
  - Timing of delivery of service, and
  - Hookup provisions.
- Method of compensation to project participants, for example,
  - Independent developer assumes development risk, operating risk, and cost of capital and charges WWTP for power taken on some discounted basis.
  - Wastewater treatment plant owner and independent developer share development costs, risks, and benefits on some basis (typically proportional to equity contribution).

6.7   Shared Savings

A common mechanism offered by energy management companies, but also available through some cogeneration equipment manufacturers, is referred to as *shared savings.* Under the shared savings approach, the energy management company finances and implements energy conservation measures at no risk to the energy user. The energy management company takes its compensation in the form of shared savings; that is, the cost of implementing these energy conservation measures plus some component for return on investment is funded by the reduced cost of utility power and fuel purchases.

Many variations to this structure may be available, including the following:

- The energy management company might be paid a fixed annual management fee.
- The energy management company's share in energy cost savings may be subject to some minimum or maximum, either in magnitude or in number of years.
- The term for sharing savings may be subject to a fixed term, with or without prospect for renewal.

For a power-generation project, a large equipment manufacturer may offer to supply either the cogeneration unit or the entire generation plant on a turnkey basis with no upfront cost to the plant owner. Compensation to the manufacturer would then be structured on the basis of a stream of payments funded by the expected savings in purchased power costs.

7.0 PROFILE OF ENERGY REQUIREMENTS

Before embarking on a project, the W/WWU owner should have an understanding of the W/WWU's overall energy requirements, condition and type of equipment (classified by process function and power requirements), and fuel resource options. As noted earlier, an owner should become aware of the energy consumption the W/WWU is experiencing by learning information on the power bill and keeping all trend plots current so there is a visual representation of the energy consumption occurring at the W/WWU. Further, an energy survey should be initiated and completed prior to any major project planning to ensure that management of the W/WWU is energy efficient and meeting all effluent requirements. At times, energy surveys will identify energy-efficient opportunities that will also result in providing additional treatment capability. However, this is site-specific—but it may be your W/WWU.

In addition, through various sources, power consumption values presented in terms of kilowatt-hours/million gallons and kilowatt-hours/pound of BOD are available. A WWTP operator can use them to compare consumption values, which will help determine if the WWTP is energy efficient. The W/WWU operator is encouraged to compare energy consumption, not cost, with these values to assess the utility's level of efficiency. This is important because the cost of energy is continually increasing; therefore, the form of measurement and comparison needs to be energy consumption and not cost.

8.0 REFERENCES

California Energy Commission, California Distributed Energy Resource Guide. http://www.energy.ca.gov/distgen/ (accessed February 2009).

Energy Services Coalition Home Page. http://www.energyservicescoalition.org (accessed February 2009).

National Association of Energy Service Companies Home Page. http://www.naesco.org (accessed February 2009).

Peterson, J. W. (2008) Personal communication with Jerald W. Peterson, Project Development Consultant, Johnson Controls, Inc.

Porter, R. (2007) Gwinnett County Department of Water Resources, Minimizing Power Costs by taking advantage of Real Time Power Pricing, Justification of Storage and Treatment Capacity to Minimize Energy Costs. *Proceedings of the 80th Annual Water Environment Federation Technical Exhibition and Conference;* San Diego, California, Oct 13–17; Water Environment Federation: Alexandria, Virginia; pp 8703–8720.

U.S. Environmental Protection Agency (1992) Energy Policy Act (EPAct) of 1992. http://www.epa.gov/radiation/yucca/enpa92.html (accessed February 2009).

9.0 SUGGESTED READINGS

Atlantic County Utilities Authority Home Page. http://www.acua.com (accessed February 2009).

Resource Dynamics Corporation (2005) Distributed Generation Technologies. http://www.distributed-generation.com/technologies.htm (accessed February 2009).

U.S. Environmental Protection Agency, Energy Star for Wastewater Plants and Drinking Water Systems. http://www.energystar.gov/index.cfm?c=water.wastewater_drinking_water (accessed February 2009).

Appendix A

Agencies and Organizations

Alliance to Save Energy

American Council for an Energy Efficient Economy (ACEEE)

American Water Works Association (AWWA)

Association of Energy Engineers (AEE)

Association of Heating, Refrigeration and Air Conditioning Engineers (ASHRAE)

Association of State Energy Research & Technology Transfer Institutions (ASERTTI)

California Energy Commission (CEC)

Center for Energy Efficiency & Renewable Energy (CEERE)

Consortium of Energy Efficiency (CEE)

Electric Power Research Institute (EPRI)

Energy Center of Wisconsin (ECW)

Florida Solar Energy Center (FSEC)

Gas Technology Institute (GTI)

Institute of Electrical and Electronics Engineers (IEEE)

Iowa Energy Center

Lawrence Berkeley National Laboratory (LBNL)

National Association of State Energy Officials (NASEO)

National Center for Energy Management and Building Technologies (NCEMBT)

National Renewable Energy Laboratory (NREL)

New York State Energy Research and Development Authority (NYSERDA)

Northwest Energy Efficiency Alliance (NEEA)

Oak Ridge National Laboratory (ORNL)

Pacific Northwest National Laboratory (PNNL)

U.S. Department of Energy (DOE)

U.S. Environmental Protection Agency (U.S. EPA)

U.S. EPA Energy Star

U.S. EPA Climate Protection Partnerships

U.S. EPA Combined Heat & Power

Water Environment Federation (WEF)

Water Environment Research Foundation (WERF)

Water Research Foundation (formerly the American Water Works Association Research Foundation (AwwaRF)

Wisconsin Focus on Energy

Appendix B

Equations for Converting from English Units to Metric Units

atm × 101.3 = kPa

Btu × 1.055 = kJ

Btu/cu ft × 37.26 = kJ/m^3

Btu/gal × 278.7 = kJ/m^3

Btu/hr × 0.293 1 = W

Btu/lb × 2.326 = kJ/kg

cfm × (4.719 × 10^{-4}) = m^3/s

cfm/ft × 1.549 = L/m · s

cfs × (2.832 × 10^{-2}) = m^3/s

cu ft × (2.832 × 10^{-2}) = m^3

cu ft/lb × (6.243 × 10^{-2}) = m^3/kg

(°F − 32) 0.555 6 = °C

gal × (3.785 × 10^{-3}) = m^3

gal × 3.785 = L

gph × (1.051 × 10^{-6}) = m^3/s

gpm × (6.308 × 10^{-5}) = m^3/s

in × 25.40 = mm

in × (2.540 × 10^{-2}) = m

ft × 0.304 8 = m

ft-lb × 1.356 = N·m

ft-lb/min × (2.259 × 10^{-2}) = W

ft-lb/sec × 1.355 = W

ft/sec × 0.304 8 = m/s

gal × (3.785 × 10^{-3}) = m^3

gpm $\times$ (6.308 $\times$ 10^{-5}) = m^3/s

gpm $\times$ 5.451 = m^3/d

hp $\times$ 745.7 = W

hp-hr $\times$ 2.685 = MJ

hp/mil gal $\times$ 0.197 0 = W/m^3

in $\times$ (2.540 $\times$ 10^{-2}) = m

in Hg $\times$ 3.377 = kPa

kWh = 3.600 MJ

kWh/d $\times$ 41.67 = W

kWh/lb $\times$ (7.936 $\times$ 10^{-3}) = MJ/kg

kWh/mil. gal $\times$ 951.1 =J/m^3

lb $\times$ 0.453 6 = kg

lb/cu ft $\times$ 16.02 = kg/m^3

lb/d $\times$ 5.250 = mg/s

lb/lb $\times$ 1 000 = g/kg

mgd $\times$ (4.383 $\times$ 10^{-2}) = m^3/s

mgd $\times$ (3.785 $\times$ 10^3) = m^3/d

mile $\times$ 1.609 = km

psi $\times$ 6 895 = Pa

scfm $\times$ (4.719 $\times$ 10^{-4}) = m^3/s

sq ft $\times$ (9.290 $\times$ 10–2) = m2

Appendix C

Estimates of Electricity Used in Wastewater Treatment

TABLE C.1   Electricity requirements for trickling filter wastewater treatment plants

	Electricity used, kWh/d[a] (except where noted)					
	4-ML/d	20-ML/d	40-ML/d	75-ML/d	190-ML/d	380-ML/d
Item	1-mgd[b]	5-mgd	10-mgd	20-mgd	50-mgd	100-mgd
Wastewater pumping	171	716	1402	2559	6030	11 818
Screens	2	2	2	3	6	11
Aerated grit removal	49	87	134	250	600	1200
Primary clarifiers	15	78	155	310	776	1551
Trickling filters[c]	352	1319	2528	4686	11 551	22 826
Secondary clarifiers	15	78	155	310	776	1551
Gravity thickening	6	15	25	37	75	138
Dissolved air flotation	na[d]	na	1805	2918	6257	11 819
Aerobic digestion	1000	2000	na	na	na	na
Anaerobic digestion	na	na	1100	2100	5000	11 000
Belt filter press	na	192	384	579	1164	2139
Chlorination	1	5	27	53	133	266
Lighting and buildings	200	400	800	1200	2000	3000
Totals	1811	4892	8517	15005	34 368	67 319
Average flowrate, mgd	1	5	10	20	50	100
Unit electricity use, kWh/mil. gal[c]	**1811**	**978**	**852**	**750**	**687**	**673**
Energy recovery (from biogas combustion)	Na	Na	2800	5600	14 000	28 000
Net consumption[f]	1811	4892	5717	9 405	20 368	39 319
Unit net electricity use, kWh/mil. gal	1811	978	572	470	407	393
UV disinfection	123	614	1229	2458	6144	12 288
Gravity belt thickener	na[4]	na	288	434	873	1604

[a] kWh/d $\times$ 41.67 = W
[b] mgd $\times$ (4.383 $\times$ 10^{-2}) = m^3/s
[c] With recirculation pumping
[d] Not applicable
[e] kWh/mil. gal $\times$ 951.1 = J/m^3
[f] Total less electricity from energy recovery

TABLE C.2   Electricity requirements for activated sludge wastewater treatment plants

| | Electricity used, kWh/d[a] (except where noted) | | | | | |
| Item | 4-ML/d | 20-ML/d | 40-ML/d | 80-ML/d | 190-ML/d | 380-ML/d |
	l-mgd[b]	5-mgd	10-mgd	20-mgd	50-mgd	100-mgd
Wastewater pumping	171	716	1 402	2559	6030	11 818
Screens	2	2	2	3	6	11
Aerated grit removal	49	87	134	250	600	1200
Primary clarifiers	15	78	155	310	776	1551
Aeration (diffused air)	532	2660	5320	10 640	26 600	53 200
Return sludge pumping	45	213	423	724	1627	3131
Secondary clarifiers	15	78	155	310	776	1551
Gravity thickening	6	15	25	37	75	138
Dissolved air flotation	na[d]	na	1805	2918	6257	11 819
Aerobic digestion	1200	2400	na	na	na	na
Anaerobic digestion	na	na	1400	2700	6500	13 000
Belt filter press	na	192	384	579	1164	2139
Chlorination	1	5	27	53	133	266
Lighting and buildings	200	400	800	1200	2000	3000
Totals	2236	6846	12 032	22 283	52 544	102 824
Average flowrate, mgd	1	5	10	20	50	100
Unit electricity use, kWh/mil. gal[d]	**2236**	**1369**	**1203**	**1114**	**1051**	**10286**
Energy recovery (from biogas combustion)	na	na	3500	7000	17 500	35 000
Net consumption[c]	2236	6848	8532	15 283	35 044	67 824
Unit net electricity use, kWh/mil. gal	2236	1369	853	764	701	678
UV disinfection	92	461	922	1843	4608	9216
Gravity belt thickener	na[c]	na	288	434	873	1604

[a] kWh/d $\times$ 41.67 = W
[b] mgd $\times$ (4.383 $\times$ 10^{-2}) = m^3/s
[c] Not applicable
[d] kWh/mil.gal $\times$ 951.1 = J/m^3

TABLE C.3   Electricity requirements for advanced wastewater treatment plants without nitrification

Item	Electricity use, kWh/d[a] (except where noted)					
	4-ML/d	20-ML/d	40-ML/d	80-ML/d	190-ML/d	380-ML/d
	l-mgd[b]	5-mgd	10-mgd	20-mgd	50-mgd	100-mgd
Wastewater pumping	171	716	1402	2559	6030	11 818
Screens	2	2	2	3	6	11
Aerated grit removal	49	87	134	250	600	1200
Primary clarifiers	15	78	155	310	776	1551
Aeration (diffused air)	532	2660	5320	10 640	26 600	53 200
Return sludge pumping	45	213	423	724	627	3131
Secondary clarifiers	15	78	155	310	776	1551
Chemical addition	80	290	552	954	2187	4159
Filter feed pumping	143	445	822	1645	3440	6712
Filtration	137	247	385	709	1679	3295
Gravity thickening	6	15	25	37	75	138
Dissolved air flotation	na[c]	na	2022	3268	7008	13 237
Aerobic digestion	1 200	2 400	na	na	na	na
Anaerobic digestion	na	na	1400	2700	6500	13 000
Belt filter press	na	228	457	689	1385	2545
Chlorination	1	5	27	53	133	266
Lighting and buildings	200	400	800	1200	2000	3000
Totals	2596	7864	14 081	26 051	60 822	118 814
Average flowrate, mgd	1	5	10	20	50	100
Unit electricity use, kWh/ mil. gal[a]	**2596**	**1573**	**1408**	**1303**	**1216**	**1188**
Energy recovery (from biogas combustion)	na	na	3 500	7000	17 500	35 000
Net consumption[c]	2596	7864	10 581	19 051	43 322	83 814
Unit net electricity use, kWh/ mil. Gal	2596	1573	1058	953	866	838
UV disinfection	77	384	768	1536	3840	7680
Gravity belt thickener	na[c]	na	343	517	1039	1909

[a] kWh/d × 41.67 = W
[b] mgd × (4.383 × 10^{-2}) = m^3/s
[c] Not applicable
[d] kWh/mil. gal × 951.1 = J/m^3
[e] Total less electricity from energy recovery

TABLE C.4   Electricity requirements for advanced wastewater treatment plants with nitrification

	Electricity use, kWh/d[a] (except where noted)					
	4-ML/d	20-ML/d	40-ML/d	80-ML/d	190-ML/d	380-ML/d
Item	l-mgd[b]	5-mgd	10-mgd	20-mgd	50-mgd	100-mgd
Wastewater pumping	171	716	1402	2559	6030	11 818
Screens	2	2	2	3	6	11
Aerated grit removal	49	87	134	250	600	1200
Primary clarifiers	15	78	155	310	776	1551
Aeration (diffused air)	532	2660	5320	10 640	26 600	53 200
Biological nitrification	346	1724	3446	6818	16 936	33 800
Return sludge pumping	54	256	508	869	1952	3757
Secondary clarifiers	15	78	155	310	776	1551
Chemical addition	80	290	552	954	2187	4159
Filter feed pumping	143	445	822	1645	3440	6712
Filtration	137	247	385	709	1679	3295
Gravity thickening	6	15	25	37	75	138
Dissolved air flotation	na[c]	na	2022	3268	7008	13 237
Aerobic digestion	1200	2400	na	na	Na	Na
Anaerobic digestion	na	na	1700	3200	7800	15 600
Belt filter press	na	228	457	689	1385	2545
Chlorination	1	5	27	53	133	266
Lighting and buildings	200	400	800	1200	2000	3000
Totals	2951	9631	17 912	33 514	79 383	155 540
Average flowrate, mgd	1	5	10	20	50	100
Unit electricity use, kWh/mil. gal[d]	**2951**	**1926**	**1791**	**1676**	**1588**	**1558**
Energy recovery (from biogas combustion)	na	na	3500	7000	17 500	35 000
Net consumption[e]	2951	9631	14 412	26 514	61 883	120 840
Unit net electricity use, kWh/mil. gal	2951	1926	1441	1326	1238	1208
UV disinfection	77	384	768	1536	3840	7680
Gravity belt thickener	na[c]	na	343	517	1039	1909

[a] kWh/d $\times$ 41.67 = W
[b] mgd $\times$ (4.383 $\times$ 10^{-2}) = m^3/s
[c] Not applicable
[d] kWh/mil. gal $\times$ 951.1 = J/m^3
[e] Total less electricity from energy recovery

Appendix D

Electricity Basics

Electricity is a condition in which electrons are displaced from atoms. Because it takes energy to displace the electrons, the displaced electrons represent energy that can do useful work. Therefore, electricity is a form of energy that results from an unbalanced atomic condition. If the electrons are merely displaced, a condition of static electricity exists. If the electrons flow back to the atoms, a condition of current electricity exists during the time of movement. Electricity can be generated in several ways, but some other form of energy is required, and the other energy form must be converted to electricity. The most commonly used methods to generate electricity are magnetic (e.g., generators and alternators), chemical (e.g., batteries), and photocells (e.g., light sensors and solar panels), with magnetic generation being the major source of artificially produced electricity in the world. Fossil fuels, hydropower, geothermal energy, nuclear energy, wind, and tidal energy are used to develop the mechanical energy to magnetically produce electricity.

In any electrical circuit, at least four factors are present: electrical pressure (voltage), electron flow or current (amperage), resistance, and power (wattage). Voltage, current, and resistance are related through Ohm's law in the following equation:

$$E = IR \qquad\qquad (D.1)$$

Where
E = electrical pressure, volts (V);
I = electrical current, amperes (A); and
R = resistance, ohms.

Ohm's law does not apply to alternating current circuits, which contain inductors or capacitors. Another relationship exists for simple circuit power P, in which

$$P = EI \qquad \text{(D.2)}$$

Where

   P = power, watts (W).

Equipment used in water and wastewater treatment plants is commonly powered by 480-V three-phase alternating current power. In this case, the equation is modified to include power factor and a multiplier (the square root of 3) to determine power. Power and power factor for three-phase circuits is discussed in detail in Chapter 3.

1.0   ALTERNATING CURRENT VOLTAGE

Alternating current voltage is determined by its effectiveness or the power it produces compared to direct current voltage. Because the voltage of each alternating current starts at zero, increases to maximum, then decreases to zero, alternating current voltage is continually varying and reversing and is never at a constant value. Therefore, an effective value has to be determined. When the voltage of an alternation reaches peak, for example 90°, it is known as *peak* or *wave-crest voltage*. The effective voltage, or working voltage, of alternating current is 0.707 times the value of the peak voltage. This value was determined by comparing alternating current voltage with direct current voltage in producing heat in a resistor. In terms of effectiveness, 1 alternating current-effective volt (0.707 of the peak alternating current voltage) is equal to 1 direct current volt. Effective voltage is also known as *root-mean-square (rms) voltage*. All common electrical instruments have readouts that are in effective or rms voltage (e.g., 120 V is the effective voltage and 170 V is the peak voltage of common alternating current).

2.0   VOLTAGE DROP

Voltage is electrical pressure. In any series electrical circuit, the total volts applied to the circuit are entirely used. No pressure returns to the generator. All amperes that leave the generator return, but volts are completely spent in the circuit. If ammeter readings are taken around a circuit, it will be found that the ampere readings are all the same. The volts spent or dropped in a part of a circuit are in direct proportion to the resistance in that part of the circuit.

Copper is the most widely used electrical conductor because of its high electrical transfer efficiency and relatively low cost compared to precious metals, which have higher transfer efficiencies. A voltage drop is present in all circuits because of resistance in the conductor. Resistance increases directly with length of wire and inversely with the diameter of the wire.

Resistance caused by improper wire sizes, poor wire connections, or rough contact points in switches results in a voltage drop of the line voltage, as well as buildup of heat and loss of power. For example, in a simple circuit with a device that requires 100 V and 5 A and has a 1-ohm resistance because of the wire, the line voltage needed will be 105 V because the 1-ohm resistance at 5 A will require 5 V of line pressure to overcome the resistance in the wire. Additionally, the power loss is represented by an equation generated by combining Eqs. D.1 and D.2, as follows:

$$P = I^2R \qquad (D.3)$$

In this example, the power loss is 25 W ($5^2 \times 1$). The 25-W power loss is energy lost in the form of heat dissipated over the length of the wire.

Regardless of the form of the resistance, heat will be produced when electricity flows through it. This is true of connections that may loosen from the constant warming and cooling of electric circuits, which causes expansion and contraction of circuit metals or of contact points in switches, which, in turn, wear from use. Loose connections, worn contact points, or other undesirable resistances can be found using infrared thermography or simple voltage measurements across the connection or switch. Infrared photographs or video cameras make heat visible, with greater temperatures producing brighter images. As discussed previously, circuit resistance results in a voltage drop. A simple voltmeter reading across a connector or switch from time to time will show whether resistance is increasing or at excessive levels.

Index